HOW TO CAST SMALL METAL AND RUBBER PARTS

2ND EDITION

WILLIAM A. CANNON

TAB Books
Division of McGraw-Hill

New York San Francisco Washington, D.C. Auckland Bogotá
Caracas Lisbon London Madrid Mexico City Milan
Montreal New Delhi San Juan Singapore
Sydney Tokyo Toronto

TAB Books is a division of McGraw-Hill, Inc.

pbk 19 20 QWF/QWF 0 7 6 5
hc 1 2 3 4 5 6 QWF/QWF 8 9 8 7

Library of Congress Cataloging-in-Publication Data

Cannon, William A.
How to cast small metal and rubber parts.

Includes index.
1. Founding. 2. Rubber goods. I. Title.
TS233.C36 1986 671.2 85-27709
ISBN 0-8306-0314-X
ISBN 0-8306-0414-6 (pbk.)

Contents

Introduction

How To Cast Small Metal and Rubber Parts-2nd Edition is written largely from the standpoint of collector car restoration, but it contains information valuable to everyone from home craftsmen and antique collectors to sculptors and inventors. **Part 1—Casting Metal Parts**—describes practical techniques for simple foundry work and tells you how you can get started in this rewarding and creative activity without spending a fortune on equipment and supplies.

Car restoration today is much more demanding than it was 20 years ago. As the hobby has become more sophisticated, there has been a trend toward greater authenticity. This demand has imposed more of a burden on restorers as original parts become scarcer. It is often difficult to find rare original parts to complete the final details of classic or antique cars. Simple metal casting has proved to be an inexpensive alternative to endless searches for parts in swap meets and flea markets. Unfortunately, finding a commercial foundry to do the "one-off" reproduction of a part can be as difficult as finding the original part itself. In such a case, the serious restorer may find it practical and interesting to do the work himself. This book tells you how.

Antique collectors are learning that broken or missing parts of antiques and art objects can be restored using simple metal casting techniques. By reproducing a part missing from a rare old brass lamp, you may convert the lamp from a nearly worthless piece of junk to a valuable antique. Artists can convert their sculptures, plaques, and wood carvings to art objects of enduring bronze, while inventors will find foundry work invaluable for producing new parts, models, and designs.

Considerable material about methods of metal finishing, grinding, and polishing parts has been added to this second edition, including suggestions for producing parts too intricate to make by conventional sand casting techniques.

In contrast to the subject of foundry work, **Part 2—Casting Rubber Parts**—draws upon more modern scientific developments in polymer chemistry. Until now, making rubber parts in complex shapes has been the exclusive province of large-scale industrial operations because it required expensive dies, equipment, and production machinery. Now comes polyurethane, a new product that you can mix and pour into simple molds where it cures to a flexible material with rubber's appearance and all of its properties. The ready availability of polyurethane brings economical production of rubber parts down to the level of the amateur craftsman and hobbyist. The potential applications are so numerous that it is practical only to describe some typical methods that you can adapt to your needs.

The commercial product Devcon Flexane comes in handy 1-pound and 10-pound kits for making substitute rubber parts. Formerly available in four hardness grades for nearly all requirements, the Devcon Flexane product line has been revised so that the material is available in only two hardness grades. A flexibilizer is required for making softer rubber formulations. This edition includes a detailed explanation of Devcon Flexane packaging and formulations revisions.

I have developed or used each operation described in this edition. The methods are practical ones that work, and they can easily be carried out by anyone with average manual dexterity.

Part 1

Casting Metal Parts

Chapter 1

Survey of Casting Methods

Casting small metal parts for restoration purposes is among the most creative and rewarding activities that the home craftsman can undertake. Relatively few amateurs have attempted this work, probably because they assume that the operations require too much equipment, skill, and know-how. Foundry work is not beyond the means or capabilities of the home craftsman, and there is no valid reason why these useful arts do not find wider application.

The production of metal castings is one of the basic processes of industry, and over the years many standardized methods have been developed. There are few other industries where successful operation depends so much on proprietary shop techniques. While numerous books have been written on foundry operations, most of them are directed toward engineering practices in high-volume production, and it is difficult for the amateur to assimilate the information published and translate it to his own needs. We shall attempt to overcome this shortcoming in Part 1.

In collector car restoration, to cite one amateur activity, many parts have become so scarce today that car owners despair of finding originals. In many cases, the shortage is due to the fact that so many car parts and trim items were originally made of zinc die casting alloy—pot metal as restorers usually call it—and this material is poorly resistant to corrosion, especially in damp urban environments and along sea coasts. Pot metal parts, if pitted or corroded, are difficult—sometimes impossible—to repair or plate.

Fig. 1-1. A few of hundreds of different reproduction parts made in my shop for antiques, coin-operated machines, orchestrions, and antique cars. All parts are made of aluminum, brass, or bronze cast in sand molds.

A lively reproduction parts business has developed for some of the more abundant makes, but for the owner of more obscure makes reproduction parts are not likely to be commercially available. There just isn't enough demand for parts in this category to make it profitable for a commercial foundry to undertake production of them.

Many small foundries will make small parts on a custom basis. If your needs are minimal this may be the best solution but production work of this kind is often fairly expensive. If you require extensive parts, or if you wish to explore this fascinating work as a creative or money-making hobby, you are encouraged to make your own parts.

Be forewarned that the kind of foundry operations that we will discuss in the following chapters is often hard, dirty work. But if you do car restoration, you are accustomed to hard, dirty work, so foundry operations should not be anything you can't cope with. Anyone with average strength and manual dexterity should be able to handle foundry work with ease.

LOW-MELTING ALLOYS

It has been suggested many times that there are low melting

alloys which can be melted on top of the kitchen stove and poured into rubber or plastic molds to make acceptable car parts. There is a variety of low melting alloys—we know of one that melts at about 150 °F. and could be melted in a double boiler on a kitchen stove. When I was in college, it was a fall ritual to raid the chemistry stock room for some of this alloy—Wood's metal, it is called—and cast up a couple of teaspoons. When a teaspoon made of Wood's metal that we slipped to an unsuspecting freshman in the dining hall disintegrated in his coffee, there was uproarious laughter and banter about the quality of dormitory coffee.

Wood's metal is expensive because it contains two costly metals—bismuth and tin. Wood's metal and similar alloys are worthless for making car parts. First, they are too soft, and second, they have the bad habit of undergoing creep at room temperature. Put another way, if they are subjected to stress at ambient temperature, the alloys will be permanently deformed. If you take a straight rod of Wood's metal, clamp one end in a vise and hang your hat on the other end it will appear to make a satisfactory hat rack, but an hour later, you will probably find the rod hanging down at an angle, and your hat on the floor.

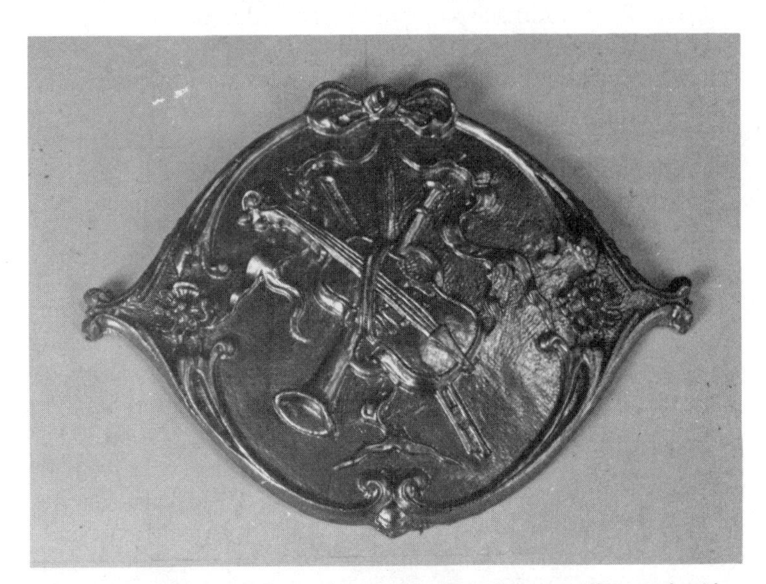

Fig. 1-2. This finely detailed reproduction of a plaque from an antique orchestrion is an example of the fine detail that can be obtained by advanced sand casting techniques. This six-inch diameter specimen is exactly as it appeared when removed from the sand mold and cleaned. It has not been retouched or polished, and even the violin strings and flower petals stand out in sharp detail.

There may be some functional car parts that could be cast from this kind of alloy, but they would have to receive almost no stress. Lead-tin, another fairly low melting alloy, is most often encountered in the form of soft solder. Solder for electrical use is usually close to 50:50 tin-lead, but various proportions are used for other applications. The most common lead-tin alloys are too soft and too weak to make useful car parts. You can make what is known as type metal when you harden type metal alloys with antimony. Lead-tin is strong and hard, but you sacrifice ductility to get hardness, so the type metals are often brittle. Most type metals expand on solidification, completely filling the molds and allowing sharp impressions when the type is used for printing.

Zinc die casting alloys, known in the trade as Zamak alloys, consist principally of zinc, aluminum, and copper. Many reproduction car parts imported from Mexico and South America are made of Zamak alloys. Without careful quality control, the properties of these alloys vary significantly. Some parts are hard, difficult to machine, and lack ductility. Zamak alloys are also difficult to weld or braze. Although primarily used for die casting, Zamak alloys can be sand cast. Since the melting temperature is fairly low (about 700-750 °F.), the furnace requirements are reduced somewhat compared to brass or bronze casting. I have never been fond of zinc alloys for making reproduction car parts because they lack ductibility and they are difficult to machine and weld.

ALLOYS PREFERRED TO PURE METALS

This is probably a good time to point out that pure metals are rarely used for making castings. The properties of pure metals usually can be improved with respect to fluidity, melting point, strength, and hardness by the addition of one or more alloying elements. The art of metallurgy has prompted the development of standard casting alloys with ideal properties for given applications. The amateur will find in the long run that using these materials yield the most favorable results. More details of the properties and applications of standard casting alloys will be presented later. By far the greatest tonnage of metal cast in industry is cast iron. Although it is deficient in strength and ductility, its low cost overcomes all other considerations in heavy machine parts manufacturing. You will probably compromise and make most experimental or reproduction parts of aluminum, brass or bronze. These alloys, while more costly than cast iron, are still reasonably inexpensive, and they have better properties in most other applica-

Fig. 1-3. This 13-inch-long plate from an antique slot machine was sand cast of aluminum alloy and retains all the fine detail of the original part.

tions. Most of Part 1 will be about casting parts in aluminum and copper-based alloys (brass and bronze) because these materials fulfill nearly all of the typical amateur foundryman's requirements for small parts fabrication.

Not all industrial processes for casting metals and alloys are adaptable to amateur production, but it is good to be familiar with them.

SAND CASTING: ECONOMICAL AND VERSATILE

Crude methods of sand casting were practiced before the dawn of recorded history and are by far the oldest casting technique. Perhaps some distant ancestor of ours picked up some copper ore at random and banked his campfire with it. The heat of the fire smelted the ore, and the molten copper ran into a crack in the ground to form a crude blade. Sand casting was invented!

Sand casting consists of pouring a molten metal or alloy into a mold of earth or sand and allowing it to solidify. Sand casting is by far the most economical way to make a few parts because molding material is cheap, plentiful, and easy to work with. It is difficult to visualize how you can make a metal part more economically than by sand casting. Machining or building up a part by welding, riveting, or brazing is almost sure to be more expensive than casting. Sand casting is a versatile process that allows you to make almost any size part.

Pure sand is not suitable for making molds because the grains lack coherence. A binder such as clay is needed to hold the sand

Fig. 1-4. Artists will find sand casting to be an easy way to reproduce their artwork in enduring metal. This six-inch plaque was reproduced from an amateur's wood carving.

particles together. Too much clay is undesirable because sand loses porosity essential for the escape of gases when the molten metal is poured into the mold.

LOST WAX CASTING FOR COMPLEX PARTS

"Lost wax" casting is called investment casting in industry. Recent excavations of Roman villages have provided excellent examples of thriving foundries that used the ancient method to make chariot parts and other useful items. During the Renaissance, the art was perfected to produce art objects.

The method consists of surrounding or "investing" a wax pattern of the object to be cast with a setting refractory cement. The mold is heated to harden the refractory and burn out the wax. The melted wax leaves a cavity in the mold that is the exact size and shape of the original pattern. The mold is filled with molten metal poured through an opening left for that purpose.

The lack of means for producing large numbers of identical wax patterns limited the industrial applications of investment casting until about 1930 when rubber dies and wax pattern machines were developed in the dental and jewelry trades. During World War II, the demand for large numbers of intricate castings spurred the development of investment casting. Today, it is used extensively for precision parts for jet aircraft engines, one of many applications. Investment casting is the most expensive foundry method. Its use is mainly for the production of complicated parts, and for pilot or experimental parts produced in small quantity.

Many car parts can be made by investment casting, but the labor costs for making a single piece may run to many times the cost of making the same part by sand casting. A typical mold for a sand casting takes only minutes to make. The same mold for an investment casting requires first that a mold of the part be made to produce the wax pattern. After this is done, the wax pattern is invested in refractory cement and the mold is heated in a furnace to burn out the wax. These operations may run several hours.

Investment casting has one big advantage over sand casting: there is no limit to the shape or complexity of the part. Sand casting is limited to less complex shapes because of the necessity of opening the mold to remove the pattern.

Using frozen mercury for patterns instead of wax is an interesting but unusual variation of investment casting. Mercury freezes at about 40 °F., so dry ice can be used to keep mercury patterns frozen. When the patterns are allowed to warm up a little after they are invested in refractory cement, the mercury melts and flows out of the molds through appropriate openings.

DIE CASTING POPULAR FOR MASS PRODUCTION

In die casting, pressure forces molten metal or alloy into the cavity of a steel mold. Alloys that can be die cast are mainly tin-base, zinc-base, aluminum-base, and to a limited extent, copper-base alloys. The finish of a die casting is very smooth. Little machining is required for die castings and dimensional accuracy is excellent. Pressures as high as 50,000 pounds per square inch are applied, so sections as thin as 0.040 inches can be cast.

A phenomenal increase in the use of die castings has occurred in the automobile industry in the past 50 years. Zinc-base alloys are used almost exclusively for carburetors, fuel pumps, radiator grilles, and trim items. Die castings account for the second largest use of zinc, and every restorer is familiar with zinc die casting alloy—commonly referred to as "pot metal." Die casting becomes economical only when 5,000 to 10,000 parts or more are produced because the steel dies used in the process are expensive.

PERMANENT MOLD CASTING

Permanent mold casting is similar to die casting except that

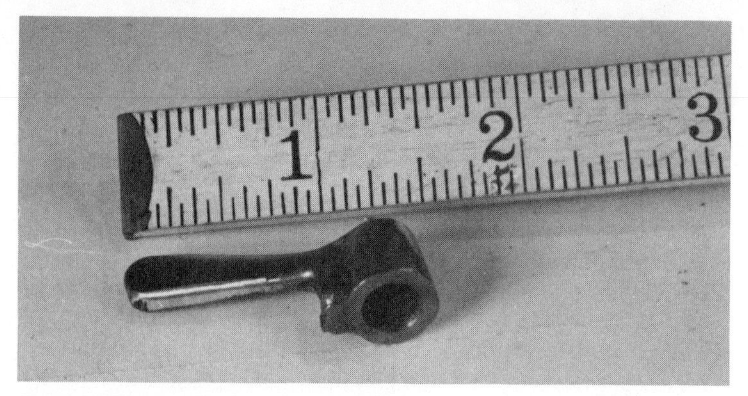

Fig. 1-5. Parts as small as this switch lever from an antique car can be readily cast, accurate in every detail.

gravity rather than pressure is used to force the metal into the die. Equipment for permanent mold casting costs less than for die casting, but the molds are still costly. Permanent mold casting is essentially a high-volume production process.

PLASTER MOLD CASTING

Plaster mold casting is a variation of sand casting, and has several industrial applications. A mixture of plaster, talc, and water is poured around a pattern. The pattern is withdrawn when the plaster has set, and the mold is baked in an oven to drive off moisture. The process is slightly more expensive than sand casting, but it claims a higher degree of dimensional accuracy.

SHELL MOLD CASTING

Shell molding is a relatively new process in the foundry industry—sometimes called the Croning Process after the inventor. It uses a mixture of sand and thermosetting resin (usually phenol-formaldehyde) to form the mold. When it touches a heated pattern, the sand-resin mixture forms a thin shell due to polymerization of the resin which binds the sand particles. The thin shell is used as a mold, backed up by loose sand or shot to give it strength. The process produces excellent finish and dimensional accuracy, and is used extensively in the auto industry to produce parts such as camshafts, rocker arms, and crankshafts. Shell molding is strictly a large-scale industrial process because expensive production machinery is required.

The amateur foundryman will be limited to sand casting, lost wax casting, or plaster mold casting. Sand casting is usually the only economical method to use when you want to make a few parts. Fortunately, it is a very versatile process and you can cast almost any size, shape, and range of metals and alloys. Two important limitations to sand casting, compared to some of the other processes are: (1) the dimensional tolerances that can be obtained are not as good; and (2) the complexity of the part to be cast is limited because it must have a "line of parting" allowing separation of the mold's two halves so that the pattern can be removed. Almost every part that was originally made by sand casting will have a line of parting.

Chapter 2

Casting Your Own Hood Ornament

The first requirement for making a part by sand casting is to have a suitable pattern. For production work, the manufacturer will invariably provide wood or metal patterns. If you are planning to make only one or two parts, the labor required to make a pattern may not be justified. In this case it is expedient to use an original part as a pattern if one is available. You can still use a broken or damaged original part if you can glue the broken parts together with epoxy resin. Missing or damaged portions can be patched or filled with resin, solder, clay, or wax.

If no original part is available, it will be necessary to make a pattern of wood using conventional wood-working or carving techniques. Firm woods such as pine are suitable, but avoid soft and weak woods like balsa because it probably will not stand up to the rigors of sand molding. A second choice is polyester auto body resin. You can form it to the approximate shape required, and after curing it you can sand and file it to the correct form. Intricate patterns in wood or resin are sure to be difficult and time-consuming to produce. Their fabrication will demand all the skills of the amateur craftsman.

SHRINKAGE AND SURFACE FINISH

Most alloys shrink when cast so that the final product will be slightly smaller than the original pattern. Shrinkage varies with the metal cast, but for brass, bronze, and aluminum alloys it averages about 3/16- to 1/4-inch per foot in all dimensions. If you were to cast a bar from a one-foot-long pattern, the part would measure about 11 3/4 inches long. Slight shrinkage is tolerable for many parts, and will not be noticed in the final piece. If dimensions are critical, then you must use an oversize pattern to compensate for shrinkage. This allowance is conventional for foundry work.

Surface finish is another problem. Most sand castings have rough and imperfect surfaces. Surface finish will be less crucial, however, if you plan to paint the part or to use it with its as-cast finish. A problem arises if you want a polished, plated, or machined surface. You must remove even more metal to get a smooth, sound surface. Fighting this two-front battle will be the subject of future discussions. First, let's make a part by sand casting.

MAKING THE SAND MOLD: FORD V-8 HOOD ORNAMENT

The following preliminary description of sand casting hopefully will arouse your interest and enthusiasm, and help you to understand the importance and significance of the fine points discussed in later chapters.

At the risk of putting the cart before the horse, the step-by-step production of a pair of hood ornaments for a Ford V-8 pickup truck will be described. The slight shrinkage we will get by using the original parts as patterns will be negligible because the parts

Fig. 2-1. The half of the flask called the drag is placed on the molding board with the patterns approximately centered.

Fig. 2-2. Molding sand is firmly rammed into the drag, compacting it around the patterns.

we're making are small (about two inches in maximum dimension), and their dimensions are not critical to the end use.

Sand molds are made in a two-piece frame called a *flask*. Metal flasks are used for production work because they must withstand considerable rough handling, but it is cheaper and more convenient for the amateur to make a wooden flask.

The patterns are placed on a flat wooden base called a *molding board*. The half of the flask called the *drag* is approximately centered as shown in Fig. 2-1. The patterns may be dusted with graphite or talc to help to separate them from the sand later on. Next, the drag containing the patterns is rammed full of molding sand as shown in Fig. 2-2. The objective is to firmly embed the patterns in compacted sand with minimum disturbance of the patterns.

PREPARING FOR THE OTHER HALF

The sand in the drag is then leveled off with the top edge using a flat board as a scraper. The drag is then inverted on the molding board, making the patterns visible. The sand around the edges of the patterns is excavated with a knife or spatula down to the "line of parting" so that the patterns can be taken out of the mold without disturbing the compacted sand shown in Fig. 2-3.

The other half of the flask or—the *cope*—is positioned on top of the drag. The parts are held together by aligning pins in the drag with matching holes in the cope. A thin layer of *parting dust* is

Fig. 2-3. The drag has been inverted on the molding board so that the patterns are now at the top. A knife blade has been used to excavate the sand down to the line of parting so that the pattern may be removed from the mold with minimum disturbance of the sand.

sprinkled onto the pattern and the surface of sand in the drag so that the halves of the mold can be separated without the sand sticking together. A brass tube called a *sprue pin* is pushed a short distance into the sand between the two patterns to form the pri-

Fig. 2-4. The cope half of the flask has been placed in position and is rammed full of sand. The tube projecting at the center is the sprue pin which forms the channel into which molten metal will be poured later.

mary channel for molten metal to flow to the mold cavities.

Now the cope half of the mold is rammed full of sand as shown in Fig. 2-4. Pneumatic rammers are used in production, but unless you are independently wealthy you will probably ram small molds like this one by hand and get a lot of free exercise.

GENTLY REMOVE THE PATTERNS

After the cope is filled with sand, and it can be mounded above the edge of the cope, the mold is separated and the pattern is removed. Gently rapping the patterns with a small mallet loosens them from the sand. If the patterns have fine detail, such as the ribs and debossed characters in our Ford ornaments, don't overdo the rapping because it may cause the sand in the fine detail to break. The patterns must be lifted out with minimum disturbance of the sand. The sprue pin is removed from between the mold halves, and a funnel-shaped opening is carved in the sand of the cope to receive the molten metal. Small channels, or *gates*, are carved in the sand between the sprue and the mold cavities. The gates will channel the metal from the sprue to the mold cavities. Figure 2-5 shows

Fig. 2-5. The halves of the mold are separated and the patterns removed. Channels have been cut in the upper half of the mold from the sprue to the mold cavities. The white appearance of the sand is due to parting dust that prevents the sand in the two halves of the mold from sticking together.

the separated halves of the mold with the gates carved in. The mold is ready to receive the poured metal when its two halves are reconnected.

POURING THE MOLD

During the mold-making process, a fired-up furnace has been

Fig. 2-6. The operator is checking the temperature of the molten charge of metal in the crucible furnace. The thermocouple has been inserted through the hole in the lid of the furnace and into the molten metal in the crucible.

Fig. 2-7. The crucible is lifted out of the furnace with tongs and the molten metal poured into the funnel-shaped opening over the sprue.

Fig. 2-8. The parts as removed from the mold are still attached to the sprue. The patterns are shown in the foreground.

melting a metal charge. In Fig. 2-6, the operator is checking the charge's temperature. A silicon bronze casting which calls for a temperature of about 2100 °F. will be made.

When the metal is at the correct temperature for pouring, the operator uses tongs to lift the crucible from the furnace and pours the metal into the funnel-shaped opening over the sprue, filling it almost to capacity as shown in Fig. 2-7.

When the casting has cooled, it is removed from the mold and cleaned. Figure 2-8 shows two perfect replicas of the patterns. The only remaining task is to cut the new parts from the sprue and polish them.

This casting is very simple. Larger castings may require more elaborate procedures, but more of this later. It took 28 minutes to make the mold, pour the metal, and remove the casting from the sand not counting the time it took to allow the casting to cool. Labor costs would be reduced by casting several parts at the same time. Up to 10 similar castings could be made simultaneously in the same mold.

Chapter 3

Alloys You Can Cast

The choice of a metal or alloy for a casting is dictated by requirements including strength, surface finish desired, ductility, and ease of machining and casting. Cast iron's low cost offsets its deficiencies, but very few reproduction car parts are made of cast iron because the materials cost usually is secondary to labor costs for limited production work. We can dismiss cast iron as a suitable candidate material for making reproduction car parts because it lacks strength and ductility, and it is difficult to polish and plate.

BRASS: DECEPTIVE AND PROBABLY UNDERSIRABLE

Brass is usually thought to be high on the list of materials for making parts by casting. Its ready availability in scrap form makes it an attractive choice for the amateur; the amateur is likely to experience some difficulties with it at first because it is hard to cast.

The term "brass" denotes a class of alloy, not a specific composition of material. By definition, brass is copper-zinc alloy containing more than 50 percent copper and may contain minor amounts of other alloying elements. The "yellow brass" of commerce ordinarily contains 65 percent copper and 35 percent zinc. It has a pleasing yellow color and polishes well. This is the brass most often used for pre-1915 car radiator shells, lamps, windshield frames, etc. It is essential to use yellow brass for the sake of appearance if you are reproducing these parts.

The generic term "brass" used without further qualification

usually means "yellow brass" containing copper and zinc. Other compositions are used for specific applications in industry. Scrap brass usually contains a variety of compositions mixed indiscriminately. You will find that scrap brass's casting properties vary considerably from melt to melt.

When brass is brought up to its melting temperature, zinc will distill out and burn above the molten alloy. This problem is alleviated somewhat by covering the melt with a protective layer of flux. But some burning of zinc is inevitable, thus changing the composition of the alloy slightly as it is melted and poured. The presence of oxidized zinc and impurities that may be present if scrap is used, causes brass to be "drossy" which frequently leads to rough surfaces on castings. The rejection rate of small castings tends to be high. The amateur foundryman will probably avoid the use of brass, particularly scrap brass, unless it is required for the sake of appearance.

BRONZE: GOOD PROPERTIES AND POPULAR

Like brass, bronze is an alloy of indefinite composition. By definition, bronze is an alloy of copper and tin that may contain small amounts of other elements. We recognize other bronzes which contain relatively little tin, such as manganese bronze, and other alloys which contain no tin, such as aluminum bronze (copper-aluminum) and silicon bronze (copper-silicon). Alloys—such as architectural bronze (copper-zinc-lead) and commercial bronze (90% copper, 10% zinc)—are called bronzes when they are brasses.

When the term "bronze" is used without qualification it usually means tin-bronze. Tin added to copper greatly increases the copper's hardness and strength. Small amounts of zinc are sometimes added to improve casting properties, and lead will improve machining qualities. Typical tin bronzes will contain about 87-90 percent copper, 6-10 percent tin, and 2-4 percent zinc. If lead is added to improve machining, usually it will be present in the amount of about one percent, and the alloy will be referred to as a leaded tin bronze. Tin bronzes as a class have good casting properties and excellent mechanical strength and ductility. Tin is a fairly costly ingredient so tin bronzes tend to be a little more expensive than yellow brasses, which contain no tin.

RED BRASS: STRONG AND WIDELY USED

Red brass is difficult to categorize because it is considered either a brass or a bronze. The composition varies, but typical red brass may contain about 85 percent copper and 5 percent each of tin, lead, and zinc. The red brasses are among the most widely used of all copper-based casting alloys and find numerous uses in the production of pipe, valves, fittings, pump housings, and plumbing fixtures. They offer an excellent combination of resistance to corrosion, high strength, and good casting properties.

MANGANESE BRONZE: NOT FOR THE BEGINNER

Manganese bronze is a favorite casting alloy for parts that require high strength and exceptional resistance to corrosion, such as ship propellers and ship fittings used in sea water. The compositions available in manganese bronze vary depending upon the strength characteristics desired, but typical compositions are around 60 percent copper and 25-36 percent zinc, with small amounts of iron, aluminum, and manganese.

Manganese bronze is difficult material to cast and you will do well to avoid it, at least until you build your skills. It is common for manganese bronze parts to be mixed with yellow brass scrap, and the contamination of the yellow brass is apt to affect the casting properties.

ALUMINUM BRONZE: DEFINITELY NOT FOR THE BEGINNER

Aluminum bronze is characterized by enormous strength which exceeds that of mild steel. It can also be heat treated to improve strength. It is extremely difficult to cast because of its narrow solidification range, and apparent high shrinkage upon solidification. The amateur is advised not to use it.

SILICON BRONZE: IDEAL FOR THE AMATEUR

The silicon bronze casting alloys are becoming popular. They have excellent casting properties, high strength which approaches that of low-carbon steel, and good corrosion resistance. Most of the silicon bronze alloys contain about 95 percent copper, 4-5 percent silicon, and minor amounts of manganese and zinc.

Of all readily available copper-based casting alloys, silicon bronze has the best combination of properties for the amateur foundryman. It is comparatively inexpensive and can be remelted repeatedly without changing composition. The as-cast surface finish is not inferior to any other alloy. Silicon bronze is available under the trade names Herculoy and Everdur.

ZINC ALLOYS HELPFUL LOW MELTING POINT

The common zinc die casting alloys can also be used for sand casting with reasonably good results. The most common alloys of this type are the Zamak alloys which contain about 95 percent zinc, 4-5 percent aluminum, and sometimes minor addition of copper and magnesium. The zinc die casting alloys have reasonably good strength and casting properties, but they lack ductility and corrosion resistance. Their only real advantage over copper-base alloys is the lower melting temperature which reduces the high temperature capabilities required for the melting furnace.

BRONWITE: DESIRABLE AND LUSTROUS

Bronwite is a patented alloy manufactured by ASARCO (formerly Federated Metals Division, American Smelting and Refining Company). It originally was developed to replace nickel silver alloys and to avoid the high cost and availability problems of nickel. Bronwite contains 59 percent copper, 20 percent zinc, 20 percent manganese, and 1 percent aluminum. It is the equivalent of nickel silver (German silver) with respect to color, corrosion resistance, strength, ductility, and hardness, but it is easier to cast because of low melting temperature (1550 °F.). Bronwite can be polished to a high luster that looks just like nickel or German Silver. Parts originally made of nickel need not be plated if made of Bronwite, and its resistance to tarnishing is a good as nickel.

You will find many applications for Bronwite in reproduction work. The relatively high shrinkage of the material upon solidification somewhat restricts its use.

ALUMINUM ALLOYS: BEST FOR YOUR FIRST TRIES

A wide variety of aluminum alloys are used for die casting and sand casting. Among the most versatile are the aluminum-silicon alloys which combine excellent castability with good corrosion resistance and high strength. Alcoa alloy 356 is often used for automotive parts such as transmission cases, cylinder blocks, oil pans, and pump bodies. It contains 93 percent aluminum and about 7 percent silicon. Sometimes minor amounts of copper, magnesium, and zinc are added to the aluminum-silicon alloys.

Aluminum alloys cast very easily and the beginner is encouraged to experiment first with aluminum casting to "get the feel of it" before graduating to the more difficult to handle brasses and bronzes. The techniques are basically the same in all cases, but aluminum alloys are much easier to handle and melt.

Scrap aluminum works quite well if you are careful to pick out only scrap castings such as gear cases, cylinder heads, and machine parts. It is best to avoid pistons and connecting rods because they are alloyed with other elements which may make the material less suitable for sand casting. Avoid wrought or structural aluminum scrap altogether, because it is not alloyed for casting purposes and will give poor results.

Scrap dealers sometimes do a poor job of segregating aluminum and magnesium alloys. Be careful not to mix magnesium with your aluminum scrap because melting magnesium may cause a fire. Once ignited, magnesium cannot be extinguished, so the only thing to do is cover it with sand to slow the burning and let it go. Never apply water or any other fire extinguishing material other than dry sand to burning magnesium.

Magnesium alloy can sometimes be identified by sight. It tends to be grayer in color and is less dense than aluminum, but these are not infallible indications. If there is any doubt, apply a drop of 1 percent silver nitrate solution to the freshly filed surface of the metal. If the metal is magnesium, a black stain will appear immediately. Aluminum and its alloys will remain unchanged. You can get a small dropper bottle of silver nitrate solution from your pharmacist.

OTHER MATERIALS FOR CASTING

The foregoing classes of alloys by no means exhaust the range of materials that can be cast. Various nickel alloys, stainless steels, monel, and magnesium alloys are routinely cast in industry. Most of these alloys present technical problems that will deter you from using them, so no further consideration will be given to them.

Precious metals such as gold and silver and their alloys are used by amateur and professional craftsmen for casting jewelry and other art objects using the lost wax method. It is a highly specialized art, and is beyond the scope of this work.

SCRAP VERSUS VIRGIN ALLOYS

Scrap brass and bronze are readily available at scrap metal dealers, usually at prices about half commercial alloy prices. You will be tempted to buy scrap metals to make castings, but this practice cannot be encouraged. Copper alloys are very difficult to iden-

Fig. 3-1. Twenty-lb. ingots of Everdur silicon bronze as obtained from a primary metal producer. Virgin alloys will usually be much superior to scrap metal for making casting because the composition is carefully controlled.

tify from appearance alone, so all kinds of alloys will end up in the same bin at the scrap dealers. You may end up with a mixture of yellow brass, aluminum bronze, and manganese bronze with uncertain quality of the finished product. If you must use scrap, select your material from a reputable dealer, and reject all scrap with an uncertain color.

For serious work, the casting alloys should be purchased in ingot form from a primary producer. More uniform results and fewer rejects will more than make up for the slightly higher raw material costs. Brass and bronze usually are sold in 25-pound ingots that are too big to melt down in typical small-scale furnace equipment. Use a power hacksaw to cut them into more convenient 4- or 5-pound chunks. Primary producers of casting alloys for the foundry trade will usually give you detailed product bulletins about the alloys they sell at no extra charge. The bulletins tell you about the alloys composition and physical properties, and give recommendations about melting practices, pouring temperatures, major uses, and the types of fluxes, degassers, or deoxidizers to be used. Always get as much information as you can about the alloys you are using and file the information for future use. Producers want the foundryman to get the best possible results, so the information supplied is almost invariably accurate and authoritative. See Fig. 3-1.

Chapter 4

Foundry Equipment: Make It Yourself

The basic equipment and materials needed to conduct foundry operations are listed below. The items are about the same as any foundry would use, except that your equipment will probably be on a smaller scale. It is helpful to visit a foundry or foundry supplier and see some of the equipment.

Equipment:
Melting furnace (crucible furnace)
Crucibles
Thermocouple pyrometer
Tongs (for handling crucibles)
Flasks (for making sand molds)
Riddles (for screening sand)
Molding boards
Trowel
Sand rammer
Sprue pins
Gloves and face shield
Skimmer

Materials:
Casting alloys
Molding sand
Parting dust
Graphite powder

Flux (for melting brass)
Degasser (for melting aluminum)

Several other minor pieces of equipment will be required from time to time, but they are things that you can find around nearly every shop or garage.

MAKING YOUR OWN $30 FURNACE

The most essential item for foundry operations is a suitable furnace for melting metals to be cast. A commercial furnace of the kind shown in Fig. 4-1 is fine for small scale operations and can be purchased complete for a few hundred dollars—more than most hobbyists want to invest. You may find a used furnace for considerably less. Avoid furnaces like the ones used for melting lead or babbitt because they will lack the required high temperature capabilities.

Fortunately, you can make a furnace for less than $30 that is almost as good as a comparably-sized commercial furnace. To make a crucible furnace, get an empty five-gallon paint can and a 50-pound bag of castable refractory, which will cost about $15-20. You can purchase castable refractory from almost any foundry supplier, and if you tell him how you plan to use it, he can probably

Fig. 4-1. A typical commercial crucible furnace for melting small quantities of brass or bronze.

recommend a good product. Be sure to get a refractory that is good for a temperature of at least 2400-2600 °F.

Refer to Fig. 4-2 to get an idea of what you will be making. Cut a couple of one-inch holes in the walls of the paint can—one about 4 inches from the bottom and the other about 2 inches from the top. It doesn't matter how the holes are spaced, but if you cut

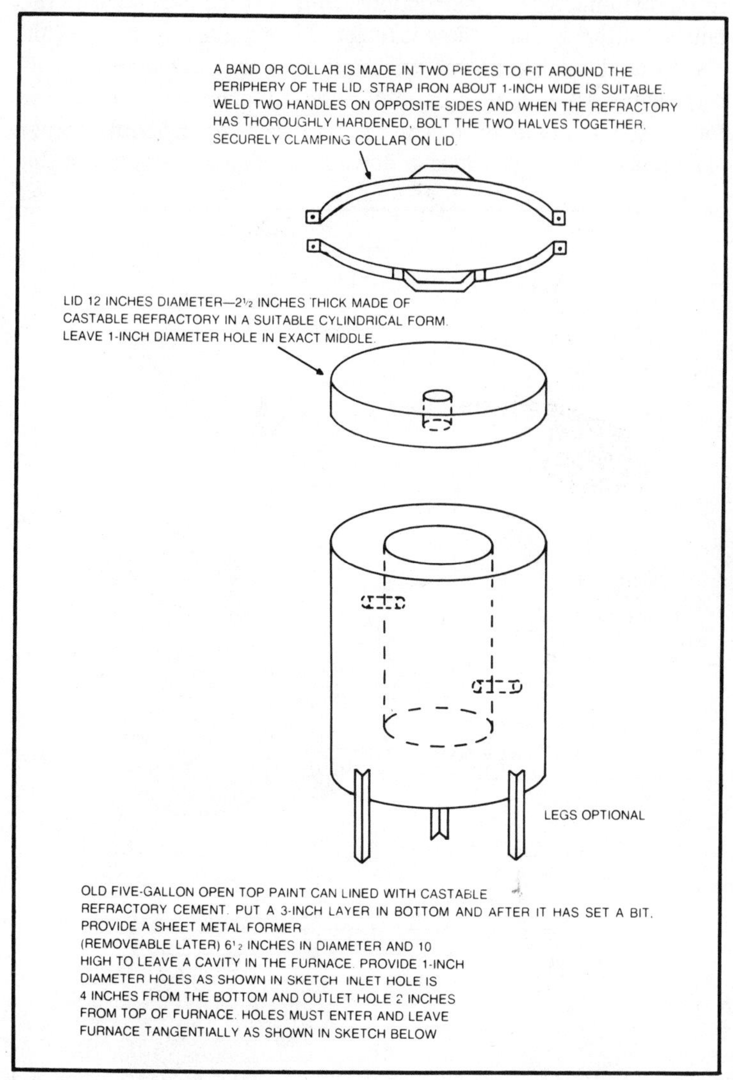

Fig. 4-2. Drawing of a small crucible furnace made from an empty 5-gallon paint can.

them roughly on opposite sides of the can, the furnace will emit the hot exit gases from the side opposite the blower. Legs can be welded on the sides of the can, but it can just as well be set on three or four bricks during use. It is a good idea to paint the exterior of the can with a good grade of heat-resistant aluminum paint.

Mix a quantity of the castable refractory with water according to the manufacturer's instructions. Pour a three-inch layer in the bottom of the can and allow it to set. Then make a former of thin sheet metal (a large tin can will do if you can find one of the correct size) 6 1/2 inches in diameter and 10 inches high. Place the former in the middle of the can so that it rests on the bottom layer of refractory. Provide tubes or dowels one-inch in diameter for the

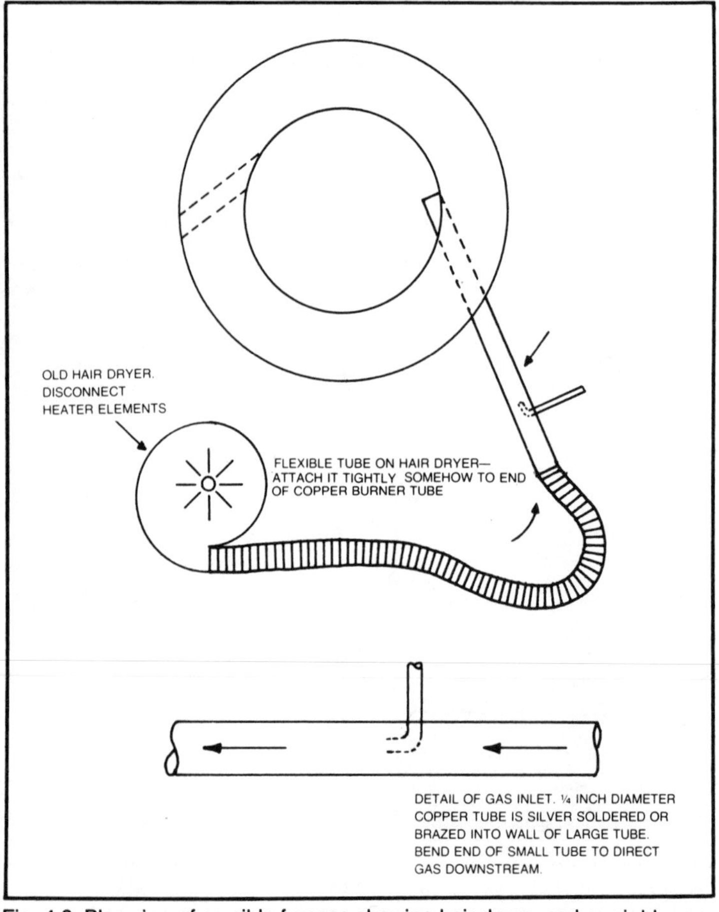

Fig. 4-3. Plan view of crucible furnace showing hair dryer used as air blower.

inlet and outlet holes. The holes must enter and leave the furnace *tangentially* as shown in the sketches of Fig. 4-3. If preferred the holes can later be drilled in the walls of the furnace with a masonry drill.

FILL WITH CEMENT AND MAKE A LID

When everything is ready, fill the annular space between the former and the five-gallon can with refractory cement, packing it in carefully a small amount at a time to eliminate voids. Trowel it carefully at the top to make it as level as possible, and allow the refractory to set for several days. In the meantime, make a lid for the furnace out of castable refractory. It should be in the shape of a circular disc, 2 1/2 inches thick and 12-inches in diameter. It can be poured into a circular mold made out of sheet metal. Leave a one-inch hole in the exact center.

Also make a collar for the lid out of a couple of pieces of iron strap about 1-inch wide. Weld or attach handles on opposite sides of the collar, as shown in Fig. 4-2, for lifting the lid off the furnace.

The recommended dimensions are only guidelines. The size of the furnace can be scaled up or down to suit individual requirements. Anything smaller than the dimensions given will restrict the amount of metal that can be melted. The size given will accommodate a No. 6 crucible which will hold up to about 13 pounds of molten brass or bronze, as much as one man can safely handle at one time. A larger crucible will require two men to lift it out of the furnace and pour safely.

THE BURNER AND THE BLOWER

The furnace is fired by a gas flame which enters the inlet hole near the bottom. The flame swirls about the inner wall of the furnace and the combustion gases escape through the exit hole and the sight hole in the lid of the furnace. An air blower is required to obtain the necessary high-flame temperatures. The air and gas are mixed in the 1-inch diameter copper inlet tube and the flame blasts out from the end of the tube inserted just inside the furnace wall.

Nearly any kind of small commercial blower can be used, but a discarded hair dryer works fine. You can usually find workable ones in second hand stores. Disconnect the heater elements, or operate it only in the "cold" position, if it has one.

Connect the gas inlet tube to a source of natural gas. Use a rubber tube for a temporary hookup. You should use copper tubing for safety in a more permanent arrangement. A valve should be placed in the gas line to regulate the amount of gas fed to the furnace. Propane can be used instead of natural gas. The amount of air can be regulated by closing off some of the air inlet holes on the hair dryer with a piece of plastic or cardboard. A damper can be placed in the air line, if necessary. Some means of independently regulating the air and gas flows to the mixing tube should be provided.

PROPER AIR-GAS MIXTURE

Take the lid off the furnace and hold a piece of lighted newspaper over the top. Slowly turn on the gas supply until it ignites and a good sized flame rises out of the furnace. Now turn on the blower. If the flame goes out, shut off the gas and try again with a little more gas flow. Some experimentation with gas and air flow regulation will be needed to get a steady flame. When properly adjusted, a bluish flame several inches long should issue from the end of the copper tube with a distinct roaring sound. If the flame tends to blow off the end of the tube, it probably means that the air flow is too great. Try again with slightly restricted air flow. Never attempt to light the furnace with the lid on; the explosion of gases may blow the lid off violently and break it.

You should operate a new furnace only a minute or two at a time periodically until the furnace refractory is thoroughly baked out. Operating it too soon may cause the furnace lining to crack or spall because of rapid steam generation. The furnace is ready for operation when its lining is thoroughly dry and no longer gives off steam when the burner is lit. You should observe the usual precautions regarding fires, explosion of unburned gases, and carbon monoxide poisoning. Operate the furnace outdoors and away from combustible materials.

A properly constructed furnace of this design should be capable of bringing a 6-pound charge of brass or bronze to the proper pouring temperature from a cold start in about 40 minutes. There are numerous other uses for this type of furnace. It can be used for forging, heat treating, and case hardening, so you might want to make one for other shop uses. A photograph of the kind of furnace described appears in Fig. 4-4.

CRUCIBLES AND PEDESTALS

The next essential item is a crucible to hold the metal while

Fig. 4-4. A homemade crucible furnace based on the designs shown in Figs. 4-2 and 4-3.

it is heated and melted. Your furnace requires one other addition—a so-called "pedestal" to place the crucible on while it is heated. It is not a good practice to place the crucible directly on the bottom of the furnace. A certain amount of debris and molten metal will inevitably fall into the furnace, and may stick to the crucible unless it is elevated about 3/4 inch on the pedestal. The pedestal may be renewed if it becomes damaged. To make a pedestal, cut discs about 3/4-inch thick by 3 inches in diameter from a fire brick, or cast them from from some of the left-over refractory cement. Sift a thin layer of pure sand over the pedestal before you put the crucible on it so that the crucible doesn't stick to the pedestal.

You must buy crucibles because they are made with highly specialized equipment. Any foundry supplier will stock them or can get them for you. Do not order a cheap fireclay crucible; they will not stand up to the rigors of repeated high-temperature firing. Products made under the Noltina trade name are leaders in the small crucibles field. Silicon carbide crucibles are much more expensive than graphite crucibles, but they are a good investment because the former outlasts graphite.

SIZES, PRICES AND CARE OF CRUCIBLES

Crucibles are available in many different sizes and shapes with capacities from a few ounces to more than 300 pounds of brass. Only the 'A' shape seems to be readily available in the small sizes, so there isn't much of a decision to make regarding the shape for small-scale work. The 2, 4, and 6 sizes are recommended for small scale operations. The dimensions are given in the table accompanying Fig. 4-5. The numbered size of a crucible indicates its approximate working capacity in kilograms of brass. A kilogram is 2.2 pounds. Fig. 4-6 is a photograph of the popular sizes of small crucibles.

The price of crucibles has skyrocketed since 1970, so it is a good idea to take good care of them. A number 4 graphite costs about $18 and the same size in silicon carbide will cost about twice as much.

Fig. 4-5. The "A"-shaped crucible is made in various sizes. The most useful sizes for small-scale work are described below.

NUMBER	HEIGHT IN.	DIAMETER TOP—IN.	DIAMETER BOTTOM-IN.	CAPACITY WATER-LB.	CAPACITY BRASS-LB
2	4½	3¾	2⅞	0.75	4.75
4	5¾	4⅝	3⅛	1.50	10.1
6	6½	5¼	3⅞	2.25	15.4

Fig. 4-6. Silicon carbide crucibles—A shape, left to right, No's 2, 4, and 6.

Store crucibles in a dry place, especially ones made of graphite. Never let them become damp by placing them on concrete floors or on the ground. Graphite crucibles should always be brought up to operating temperature slowly. Never put a cold crucible in a hot furnace or turn the burner flame directly onto a cold crucible. Fire up the furnace for a few minutes, then shut it off and place the crucible in it so that it warms gradually from the furnace's residual heat. Alternatively, the cold crucible can be inverted over the sight hole of the furnace if it is already operating. This procedure will gradually warm and dry the crucible. These precautions aren't necessary with silicon carbide crucibles because they do not absorb moisture and are more resistant to thermal shock.

Do not jam a lot of cold chunks of metal into a crucible before firing it up. Expansion of the metal may crack the crucible. Pile in the pieces loosely with plenty of room for expansion. Never let a molten charge of metal solidify in a crucible; pour out substantially all of the melted charge before it solidifies. Remelting of a solidified charge may also cause the crucible to crack from expansion of the metal.

Do not use a crucible with any signs of cracking. It may break while lifting or pouring, releasing its charge of molten metal with disastrous effects on the furnace or, worse, on nearby personnel.

Lifting the crucible out of the furnace and pouring the molten metal is the most critical and dangerous operation in foundry work. The worker must wear heavy leather or asbestos gloves, leather shoes, and a face shield. Preferably, he should also wear heavy leather or asbestos leggings. Foundry suppliers offer a variety of this type safety equipment, and it is the best investment that can be made.

Never fill a crucible excessively. In no case should the level of molten metal exceed about four/fifths of the maximum capacity of the crucible. Use only special tongs to lift and pour. Never use makeshift equipment.

MAKING YOUR OWN TONGS

Crucible tongs are made in several designs and styles and can be purchased from foundry suppliers, or you can make your own based on the designs shown in Fig. 4-7. The simple ones shown in Fig. 4-8 were made from scrap metal. A different size tong is required for each crucible so that the tong's arc-shaped ends fit each crucible's contour firmly and securely. The L-shaped style shown in Fig. 4-7 seems to offer a safety advantage but it is more difficult to construct.

JUDGING VERSUS MEASURING TEMPERATURES

Many amateurs will be tempted to judge the proper pouring temperatures of their metals or alloys by the melt's appearance alone. This procedure can be used, but it is not practical. If one melted the same alloy day after day, he might become skillful enough to judge his temperatures precisely. Otherwise, it is ex-

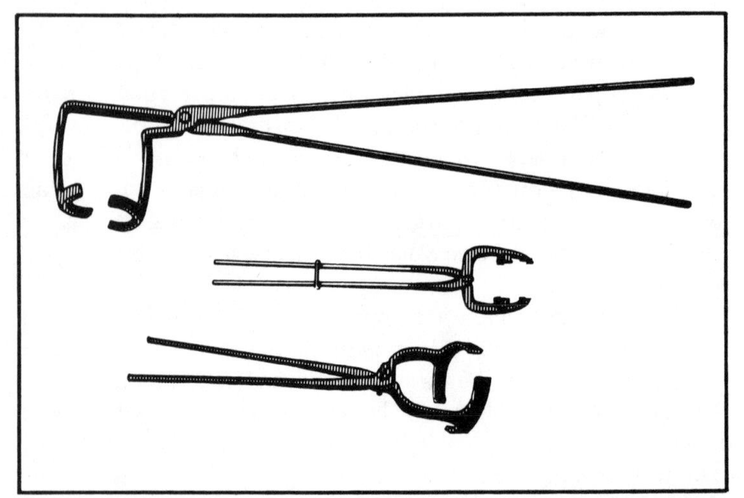

Fig. 4-7. Examples of commericial tongs for lifting and handling small crucibles.

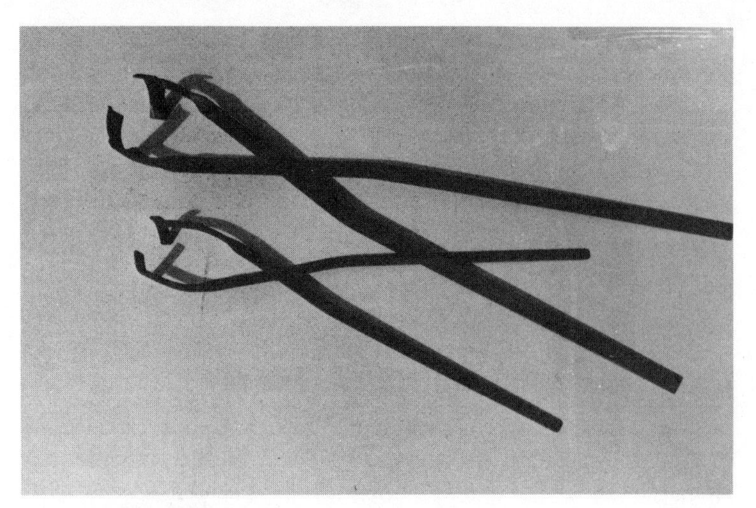

Fig. 4-8. Homemade tongs of a practical design.

tremely difficult to rely solely on visual judgment.

You can estimate the approximate temperature inside the furnace or crucible by looking through the sight hole in the lid and comparing the color with this scale:

Color	Temperature °F.
Faint red	900
Blood red	1050
Dark cherry red	1150
Medium cherry red	1250
Cherry red	1450
Bright red	1550
Orange	1700
Yellow	1850
Light yellow	2100
White	2300

The big problem with this table is the subjective interpretation of color. The amount of illumination is a factor, too. What looks like bright red in a dark shop corner or on a dark day may look like dull red in a brightly lit environment. The old time blacksmith had to judge his colors accurately by sight alone, but we can rely on more scientific methods.

So how does one measure the temperature of molten metals—far above the thermometer's range? One instrument used is the optical pyrometer, but it costs several hundred dollars so it is beyond the reach of typical amateurs. The optical pyrometer depends on comparing and matching the color of a glowing filament with the color of the object being measured. The observer adjusts the electrical current through the filament until its image disappears when viewed against the background of the furnace or crucible interior. Presumably the filament and object are then at the same temperature, and it can be read from a calibrated dial on the instrument. Readings can be taken very rapidly and at some distance from the furnace.

MAKING A THERMOCOUPLE PYROMETER

The next choice for high temperature measurement is the thermocouple pyrometer. Its operation depends on the principle that when the junction of two dissimilar metals is heated, an electrical voltage (or current) is generated that can be measured in an external circuit by a device such as a meter or potentiometer. The voltages (or currents) generated are quite small but are readily measurable with simple equipment.

Basically, all that is necessary is to connect a thermocouple—the two pieces of dissimilar metal in the form of a pair of wires—to a simple direct current meter. The current, and hence the meter reading, depend on the temperature of the junction and on the resistance of the meter and wires, so a thermocouple does not measure temperature directly. It can be calibrated by comparing the reading at known temperatures.

Obtain a 0-50 milliamp direct current meter from an electronic or surplus store. It should have a scale length of at least 2″ for accuracy and should be readable to one milliamp or better.

Next obtain a chromel-alumel thermocouple made of 8 to 14 gauge wire, equipped with ceramic insulators, with a minimum length of at least 18 inches. Connect the two leads of the thermocouple to the meter through intermediate heavy copper or brass wire (see Fig. 4-9). The connection from the thermocouple wire to the lead wires should preferably be brazed or silver soldered, but good mechanical connectors should work. Tight, secure electrical connections throughout are essential for stable readings of the instrument. The lengths of the intermediate lead wires can be made to suit, but about 18-24 inches will suffice for readings in a small furnace.

Now the thermocouple pyrometer is ready for calibration. If the meter needle deflects downscale when the tip of the thermocou-

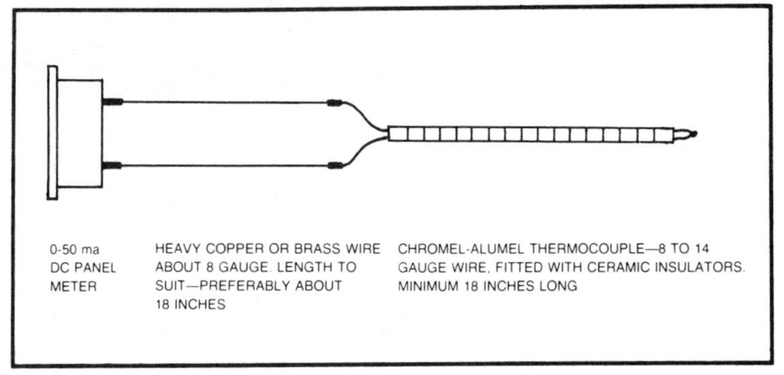

Fig. 4-9. Schematic of a simple thermocouple pyrometer.

ple is heated (you can use a match), reverse the leads to the meter. The meter should read at or near zero when the thermocouple is at room temperature. If it does not, adjust it to zero with the small screw below the meter face.

CALIBRATING THE PYROMETER

If you are fortunate enough to have available a laboratory furnace or a heat treating furnace equipped with calibrated temperature controls, it is a simple matter to insert the tip of the thermocouple into the furnace at known temperatures and calibrate the readings of the meter accordingly. Lacking a calibrated furnace, it is still possible to calibrate by using the known melting points of certain metals. For this purpose, use pure aluminum (melting point 1220 °F.) yellow brass (melting point 1700 °F.), and pure copper (melting point 1975 °F.).

Obtain a few pounds of each metal from a scrap dealer and melt them individually in crucibles. Bring the charge to slightly above the melting point, then shut the furnace off and let the charge cool until it just begins to show signs of solidifying around the walls of the crucible. Insert the thermocouple tip a couple of inches into the metal, allow the meter readings to stabilize, and record the exact scale reading. Repeat the operation for the other two metals. Plot the scale reading obtained against the known melting points to get a calibration curve similar to the one shown in Fig. 4-10.

If you don't want to make the calibration as suggested, the thermocouple pyrometer can still be used without any precise calibration. If a certain meter reading gives a satisfactory casting with a given metal, the same temperature can be reproduced later by bringing the next charge to the same meter reading. A little ex-

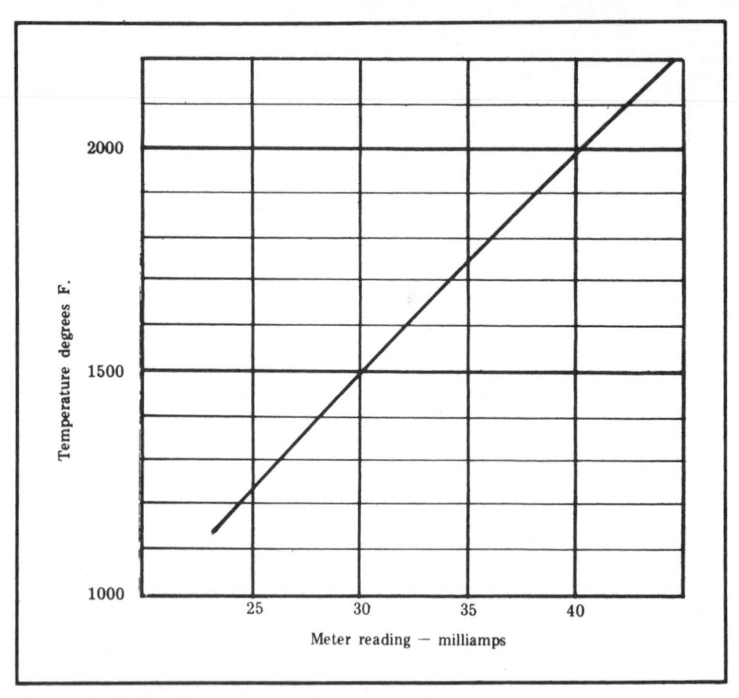

Fig. 4-10. Calibration curve for thermocouple pyrometer.

Fig. 4-11. Workbench suitable for making sand molds. The table top must be smooth and sturdy enough to withstand the impact of sand ramming.

Fig. 4-12. A collection of tools for sand molding: brushes, trowels for handling sand, and spatulas and knives for carving and excavating sand. The bulb is for blowing sand particles from mold cavities.

perimentation will soon enable the operator to produce good results from melt to melt.

You can buy thermocouple pyrometers complete and calibrated ready for use but you can make your own for a fraction of the cost. Thermocouples can be supplied by manufacturers of furnaces and similar equipment. See the Appendix for mail order supplies. Use only a chromel-alumel couple; others lack the high temperature requirements for this work.

It is a good practice not to immerse the thermocouple tip in the molten metal any longer than necessary to get a stable meter reading. Prolonged immersion of the thermocouple shortens its life.

OTHER EQUIPMENT NEEDED

You will also need a sturdy workbench for molding. A table about 2 feet wide × 6 feet long will do, and it should be covered with a continuous sheet of 3/4-inch plywood. Boards should be nailed along the back and sides of the bench top to keep sand from falling off the table. Figure 4-11 shows a practical molding bench.

A few hand tools including brushes, trowels, spatulas, and knife blades should be reserved for molding work. A collection of tools is shown in Fig. 4-12

Chapter 5

Molding Sands, Fluxes, Degassers, and Flasks

The typical foundry molding sand resembles ordinary sand. It usually is screened to a definite size fraction, and mixed with 10—20 percent clay to form a binder to hold the sand grains together. Molding sand ordinarily is quite inexpensive. In many areas, you can dig sand out of the ground with the correct sand/clay proportion. Few natural sands possess the desired properties in all proportions, however, so usually other ingredients are added to improve them.

GOOD MOLDING-SAND CHARACTERISTICS

Molding sand must have several characteristics. First, it must be cohesive so that the individual grains stick together while the pattern is being removed, otherwise the mold would break apart. Second, it must be porous enough so that gases and water vapor can escape when molten metal is poured into the mold. To a certain degree, the properties of cohesion and porosity work at cross purposes. Adding clay or clayey material will improve the cohesiveness of the sand grains, but will tend to reduce porosity. Molding sand must also be refractory to withstand the molten metal's high temperature.

The size and shape of the sand grains influence the properties of the molding sands. Rounded grains such as beach sands do not stick together nearly as well as sharp, irregular grains which interlock and provide a stronger structure when rammed into the mold. Sharp grained sands, therefore, are preferable because less clay is required for bonding and the sands are more porous.

Your local foundry supplier should be able to furnish clay-bonded molding sands in 100-pound bags. This type of molding sand is usually shipped dry. It is necessary to "condition" them before use by adding water until the sand develops the correct adhesion. The conditioning process is best carried out by sprinkling the sand with water while it is turned over with a trowel or shovel. Add enough water so that a handful of the sand compressed by clenching your fingers will stick together like a snowball, leaving a distinct pattern of the fingers and lines in the palm. Excessive water should be avoided. This kind of molding sand can be reused and conditioned when necessary, but will eventually "wear out" and should then be discarded.

PETRO BOND MOLDING SANDS

Although traditional molding sands are adequate for most routine foundry work, you will achieve far superior results with small parts if you use a synthetic molding sand called prepared Petro Bond, a registered trade mark of the National Lead Company. Petro Bond is a waterless mixture of very fine sand, oil, Petro Bond bonding agent, and a catalyst. Combining extremely fine sand with oil instead of water yields surface finish and detail in castings that you cannot achieve using conventional water-bonded sand. Surface finish on automotive trim parts is all important because you have to prepare many of the parts for plating. Smooth, finely detailed castings save grinding and polishing—tasks that are often more expensive and time consuming than making the castings. The lack of water in Petro Bond reduces gas generation. *Never* add water to Petro Bond.

For the amateur foundryman, an often discouraging activity when conventional molding sands are used becomes a real pleasure when you realize superior results with Petro Bond prepared sand. Prepared Petro Bond sand can be re-used except for the thin layer of sand in direct contact with the casting. This layer will be burned black and should be separated from the rest of the sand and discarded. The blackened sand can be reconstituted, but the effort is hardly worth the trouble for small-scale operations.

Prepared Petro Bond sand is available from selected foundry suppliers in principal cities. It costs several times as much as conventional molding sands, but it is worth the extra cost. Petro Bond suppliers are listed in the Appendix.

GRAPHITE POWDER AND PARTING DUST

A material called *parting dust* makes it easier to separate the two halves of the mold and prevents the sand from sticking to the patterns. Your foundry supplier will stock this material. You can buy a liquid parting compound, but it is not recommended for use with Petro Bond sands. About 10—25 pounds of parting dust will last for years, so avoid buying too much. Unfortunately, the minimum package amounts of some foundry products are more than the typical amateur needs. A sympathetic supplier will "break" a package to sell smaller-than-minimum shipping quantities. Also buy some *graphite powder*, a soft black carbon that is dusted onto patterns to keep sand from sticking to them. A pound or two will go a long way.

Stockmen or warehousemen around foundry suppliers often have been foundry workers at one time. If one takes an interest in what you are doing, he may offer some helpful advice about what materials to buy.

The once common practice of using fine sand as parting dust is not recommended because it can cause silicosis—a condition of massive fibrosis of the lungs marked by shortness of breath caused by prolonged inhalation of silica dusts.

FLUXES, DEOXIDIZERS, AND DEGASSERS

It is almost essential to use a flux to melt brass. It forms a molten layer over the surface of the brass and prevents access of oxygen from the air that will burn the zinc and cause excessive amounts of dross to form on the melt. Your foundry supplier will sell brass melting flux which consists primarily of borax. Do not attempt to use the regular household type of borax for this purpose because it contains water which can lead to serious trouble.

Silicon bronze, which was mentioned earlier as an ideal casting alloy for amateur use, can be melted without a flux and none is recommended. A small amount of the silicon in that alloy will be oxidized when the alloy is melted. The silicon oxide that forms produces a layer on the surface of the melt preventing further oxidation.

Other copper alloys, such as tin bronze or commercial bronze, may not require fluxes when melting. The manufacturer of the alloy can usually provide guidelines for the use of fluxes with his product. If you use a flux, add a tablespoon to the crucible before the metal melts.

Some copper alloys, notably tin bronzes and red brass, benefit where a deoxidizer is added to the molten metal. Phosphorus copper is the deoxidizer almost universally used in foundry practice today. Your casting metal supplier should be able to furnish phosphorus copper in shot or granular form. Buy only a minimum amount because you will use very little. A small amount of phosphorus copper is added to the molten metal immediately before pouring (0.1 ounces per five pounds of metal). You should accurately weigh the phosphorus copper you add, if possible. Otherwise add a piece about the size of a buckshot, or its equivalent in granular form, to each five pounds of metal you plan to pour.

The deoxidizer removes oxygen from the molten metal, increases its fluidity, and by eliminating dissolving gases reduces porosity in castings. You should use deoxidizers with copper alloys containing lead or tin, but it is not necessary to use them with brass, silicon bronze, aluminum bronze, or manganese bronze.

It is not necessary to use a flux to melt scrap aluminum, but a degasser is strongly recommended. Aluminum degasser briquets (hexachlorethane) are the most convenient form of degasser you can use. Add a chunk the size of a sugar cube to the aluminum after you melt it. Use a flat piece of steel bent at the end to hold the chunk under the surface of the molten aluminum for a few seconds. (Important: Use a face shield and gloves to protect yourself because the reaction may be violent.) You should degas the aluminum outdoors, or in a room with good ventilation to avoid breathing the chlorine fumes that are liberated. In a few seconds the aluminum will be rid of undesirable gases and some of the metallic impurities will oxidize. Skim the surface of the aluminum with a spoon-shaped steel utensil, and degas again, if necessary. Aluminum that has been degassed before pouring will produce much sounder castings, free of bubbles and inclusions. Never use the aluminum degasser for any other metal or alloy.

FLASKS: FRAMES FOR MOLDS

In the foundry trade a flask is a frame that holds molded sand. Small flasks used for production work are ordinarily made of metal for strength and durability, but you will find it more economical to make wooden flasks.

You'll need a wide variety of flasks depending upon the size, shape, and configuration of castings you'll want to make. Let us start by describing how to make a simple flask that will be useful

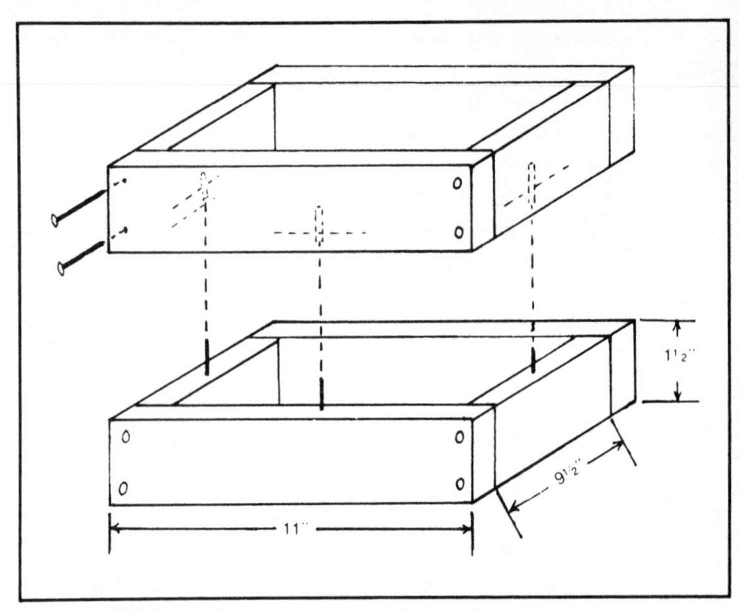

Fig. 5-1. Suggested design for a "starter" flask made of strips of 1 by 2 pine lumber. Dimensions may be varied to suit individual requirements.

for molding many small parts. Other sizes you can make by modifying the basic dimensions given.

MAKING YOUR OWN FLASK

You can use any good straight-grained lumber to make flasks; white pine is good because it is easy to work with and readily nailed without splitting. For a starter, get an eight-foot-long one by two.

Fig. 5-2. The wood strips, cut to length and carefully squared off at the ends, are ready to nail together.

A finished 1 by 2 should have a cross-section close to 3/4 by 1 1/2 inches. The design sketch in Fig. 5-1 is based on this assumption. Cut four pieces from the 1 by 2 precisely 11 inches long and four pieces 9 1/2 inches long. Make the cuts perfectly square on the ends, preferably with a miter box or circular saw as shown in Fig. 5-2.

Assemble the pieces as shown in Fig. 5-3, nailing them at the corners to form two identical frames or boxes. To hold the boxes in perfect alignment, place three locating pins in the edges of one of the frames, then match them with holes you have drilled in the other frame. Nails—about ten penny size—make good pins for this purpose when the heads are cut off. Ideally, the holes should be tight enough to hold the two frames in precise alignment, but loose enough so that the frames can be readily separated.

TELLING DRAG FROM COPE (AND A RIDDLE)

The two halves of the basic two-part flask are called the *drag* and the *cope*. The bottom part—the part with the pins permanently imbedded—is the drag and the top half is the cope. You can always remember which is which by thinking that to "drag" something requires it to be on the floor or ground, while a "cope" or "coping" is a covering.

You can profitably add some embellishments to your handiwork later. You can use screws instead of nails, and make corner joints

Fig. 5-3. The completed flask based on the design given in Fig. 5-1.

Fig. 5-4. A collection of flasks of different sizes for molding various parts.

more secure with metal corner brackets. You can also use devices to hold the two halves of the flask together. Small trunk latches are useful for this purpose. Using a method to hold the cope and drag together will prevent the possible separation of the mold

Fig. 5-5. The flask shown in this view is about 15 by 15 inches. This size flask will require that slats be inserted in the cope to support the sand so that the sand does not fall out of its own weight when the cope is lifted off the drag to remove the pattern.

caused by the cope's tendency to float when the molten metal is poured in. You will develop your own ideas for improvements as time goes on.

Figure 5-4 shows a collection of flasks for making various molds. Flasks more than about 10 inches across should be fitted with slats in the cope half to help hold the sand and prevent it from falling out when the cope if lifted off the drag, as shown in Fig. 5-5.

You must have one other item before you start making a mold. You need a riddle, or a "sand sifter," to screen lumps from the sand that makes contact with the pattern. You can save money by building your own riddle while you make the flask described above. Just tack some ordinary window screen over one side of an extra frame that you can make that's similar to the cope and drag.

Chapter 6

How to Make and Pour Molds

A trowel for handling sand (a garden variety will do), a pocket knife, a spatula, a couple of 1-inch-wide paint brushes, and a hand rammer are some of the other tools you'll need for serious mold making.

A typical hand-bench rammer for foundry work is made of hard wood, cylindrical on one end and wedge shaped on the other with a hand grip in the middle. Most of them are too large for the type of small scale work we are discussing, so a brass or steel cylindrical bar about one and one-half-inches in diameter and five or six inches long will work nicely. It is a good idea to have another rammer for working in narrow places. You can use a six-inch-long rectangular brass or steel bar 1/2-by-2-inch cross section.

The first step is to place the drag on the molding board as shown in Fig. 6-1 with the locating pins pointing downward. Drill holes in the molding board for the pins to fit into. Lay the pattern (shaded) onto the molding board approximately centered in the drag. In some cases it may be better to slightly offset the pattern in the drag so that the sprue is nearer the center of the drag. It is not necessary in this case because the pattern is small. The molding board can be any flat board slightly larger than the drag. Plywood about 3/4 to 1 inch thick will serve very nicely.

RAMMING THE MOLDING SAND

Sprinkle some parting dust or graphite powder on the pattern. Use the riddle to sift some molding sand over the pattern until it

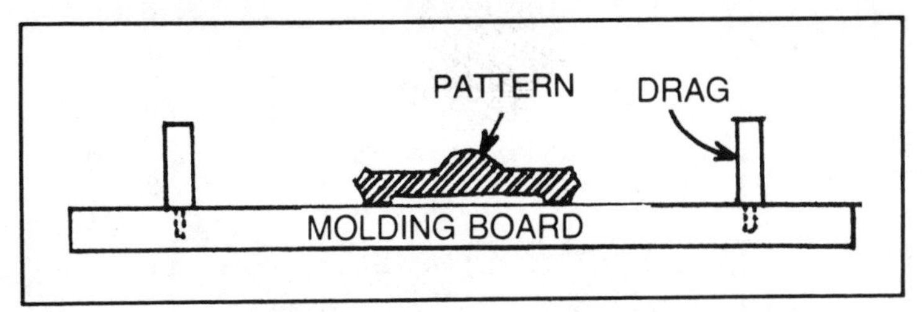

Fig. 6-1. Step 1: Lay the pattern (shaded) on the molding board and place the drag in position. The molding board can be any flat board slightly larger than the drag. It must have suitable holes in it for the locating pins.

is covered. Ram more sand in place over and around the pattern, eventually filling the drag with rammed sand above the top edge. Use a straight piece of board to strike off the sand level with the top of the drag as shown in Fig. 6-2.

It takes a little practice to do a good ramming job. First, the rammer should never strike the pattern forcibly because it may damage it or dislodge the pattern from its position. It is a good practice to press sand around the pattern with the fingers until a firm layer is built up around it. Ramming must be firm enough to consolidate the sand, but not hard enough to reduce the porosity which may prevent gases from escaping when the mold is poured. It is important that ramming is uniform throughout the mold. Any soft spots left may lead to distortion of the casting. Ramming is less critical if Petro Bond sand is used because less gas evolution occurs, so Petro Bond sand mixes can be rammed hard without problems arising from loss of porosity.

PARTING LINE AND SPRUE PIN

Invert the drag on the molding board by picking it up and

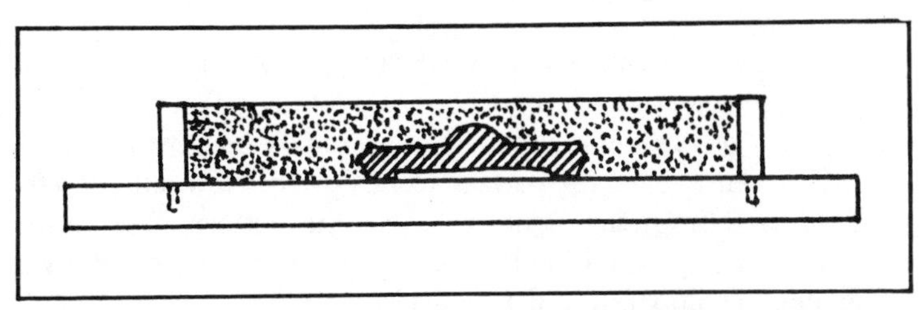

Fig. 6-2. Step 2: Sprinkle parting dust or powdered graphite on the pattern. Use a riddle to sift enough molding sand on the pattern to cover it. Add more molding sand and tamp it firmly over and around the pattern, filling the drag slightly above the top edge. Use a straight board to strike the sand off level with the top of the drag.

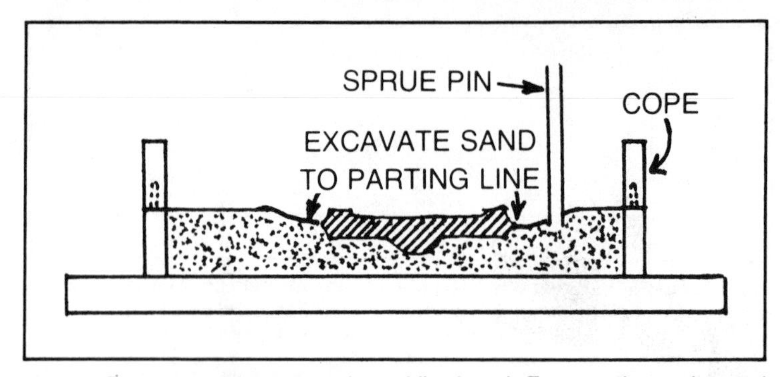

Fig. 6-3. Step 3: Invert the drag on the molding board. Excavate the sand around the edges of the pattern down to a suitable line of parting on the pattern. Place a sprue pin in the sand adjacent to the pattern. The sprue pin may be 1/2-inch diameter brass or copper tube. Position the cope on top of the drag.

turning it over. Use a knife blade and spatula to excavate the sand around the pattern down to a line of parting. On a complicated pattern, you may want to study the pattern carefully beforehand so there will be no problem finding the line of parting. On a simple pattern of the kind depicted in Fig. 6-3, the line of parting is fairly obvious; it is about half way down on the edge of the flange all around the circular pattern.

Place a sprue pin (1/2-inch diameter brass or steel tube) in the sand adjacent to the pattern and about one inch away from it. The location of the sprue is not important for symmetrical patterns because the metal can be fed in anywhere around the edge with equally good results. For more complicated patterns, the location of the sprue may require some judicious study for best results. We will get into that a little later in this chapter. If the pattern has some thin sections which are hard to fill completely, two or more sprues may actually be required to feed metal to several locations in the mold.

Place the cope into position on the drag. Sprinkle parting dust on the sand and the exposed pattern portions. You can apply parting dust by placing it into a cloth bag (an old sock will do) and shaking it gently over the surface to be dusted. Now sift enough molding sand from the riddle to cover the pattern. Add more molding sand and ram it in place, mounding it above the top edge of the cope as illustrated in Fig. 6-4.

FINAL PREPARATIONS (AND ELABORATIONS)

Carefully lift the cope off the drag and set it aside. Stand it on

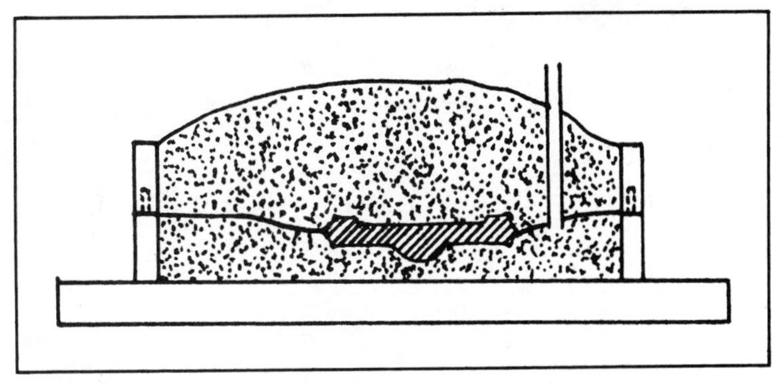

Fig. 6-4. Step 4: Sprinkle parting dust on the sand and pattern. Parting dust can be applied conveniently by placing it in a cloth bag (an old sock will work). Sift enough molding sand on top of the pattern to cover it. Now add more molding sand and ram it into place, mounding it above the top of the cope.

edge or lean it against something so that none of the sand is disturbed. Cut a channel in the sand from the sprue to the pattern edge. This gate may be v-shaped or rectangular with a cross-sectional area somewhat smaller than that of the sprue. Tap the pattern gently so that it is loosened from the sand. Carefully lift the pattern out of the drag, disturbing the sand as little as possible. Blow out any loose sand in the mold. Pull the sprue pin out of the cope and cut a funnel-shaped opening in the sand in the cope. Replace the cope in position on top of the drag. The mold is ready to pour. The completed cross-section of the mold is shown in Fig. 6-5.

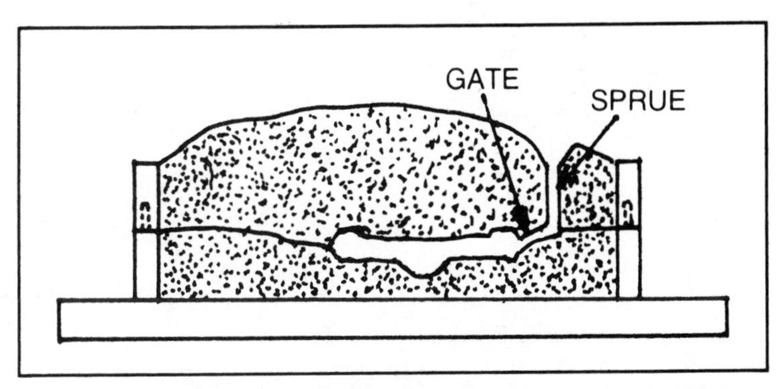

Fig. 6-5. Step 5: Carefully remove the cope from the drag and set it aside temporarily. Pull out the sprue pin and cut a funnel-shaped opening in the sand in the cope. Cut a channel (gate) in the sand extending from the sprue to the pattern. Tap the pattern very gently until it is loosened from the sand and carefully remove it, disturbing the sand as little as possible. Replace the cope on top of the drag, and the mold is ready to pour.

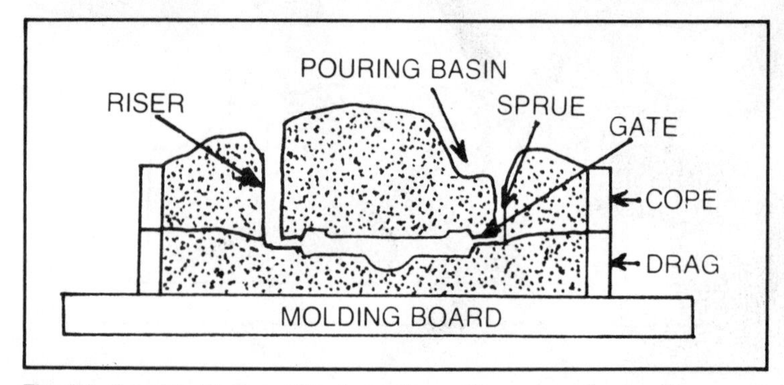

Fig. 6-6. A mold with riser. The riser will provide a reservoir of molten metal to feed back to the casting as the casting cools and shrinks, thereby tending to eliminate shrinkage cracks or voids in the finished casting. Risers are especially important if the casting has heavy sections.

The molding operation illustrated in this chapter involves a simple part. Large parts may require *risers* to prevent shrinkage cavities in the part when it is cast. Risers serve two useful purposes: they carry off sand or slag from the mold and they feed metal back to the mold as the part solidifies and shrinks. A typical mold with a riser is shown in Fig. 6-6.

If you want to make a casting with two relatively thick sections joined by a thin section, it may be impossible to feed metal into only one end of the mold. By the time the first thick section filled, and started to fill the other end through the narrow open-

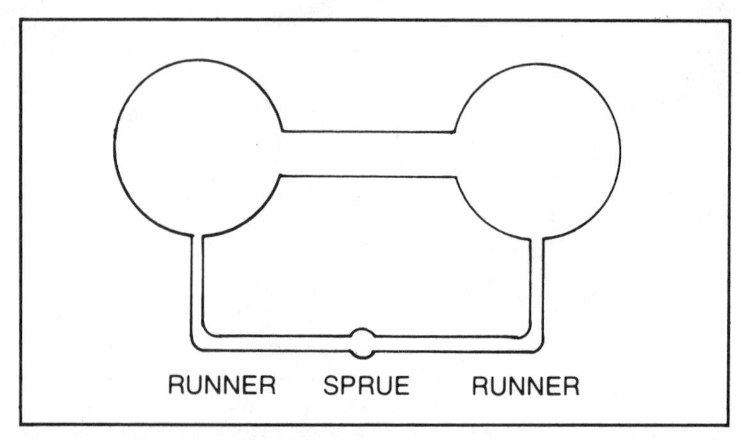

Fig. 6-7. A casting consisting of two heavy sections joined by a smaller diameter, dumbbell-shaped section is fed preferably from both ends using a runner to convey metal from the sprue to each end. The sketch is a plan view looking down on the casting from above.

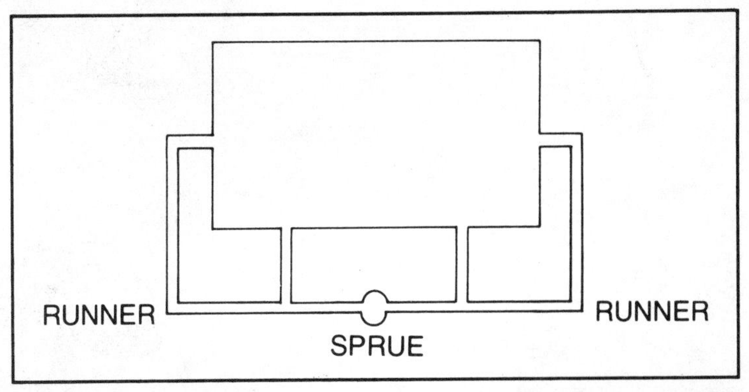

Fig. 6-8. A thin, flat casting such as a bas-relief plaque should be fed from several points around the periphery of the casting using a system of runners as shown in this plan view.

ing, there is a good chance that the metal would be cooled sufficiently to solidify in the neck. In this case, *runners* should be cut into the sand of the mold to carry metal from the sprue to fill both ends of the mold simultaneously. The scheme is shown in Fig. 6-7.

Large, thin castings should also be fed simultaneously from several positions around the edge, using runners as shown in Fig. 6-8.

The steps involved in molding a Trippe light bracket are recorded photographically in Figs. 6-9 through 6-14.

Fig. 6-9. The pattern, an original part, is laid on the molding board approximately centered in the drag. The holes in the part have been partially filled with clay to prevent unpredictable sand separation at these points during the molding process.

Fig. 6-10. After sand has been rammed around the pattern and struck off level with the drag, the drag is inverted on the molding board bringing the pattern to the top.

Fig. 6-11. Sand around the pattern is excavated down to the line of parting.

Fig. 6-12. The cope has been positioned on the drag, a sprue pin has been inserted in the sand about 1 inch from the pattern, and the surfaces are sprinkled with parting dust.

Fig. 6-13. After the sand is rammed in the cope, it is lifted off the drag.

Fig. 6-14. A gate is carved from the sprue to the pattern, and the latter is carefully lifted out of the mold. A pouring basin is carved in the sand near the top of the sprue after the pin is lifted out of the mold. When the cope is returned to its original position, the mold is complete and ready to pour.

POURING THE MOLD

When the mold is ready, the next step is to pour in the metal. Usually the operator will melt his charge of metal while he prepares the mold. Pouring the metal is probably the most critical part of sand casting, and it is the most difficult to describe adequately. The foundryman who intends to do good work will have to experiment with his pouring temperature until he gets some experience with different kinds of castings.

The first thing the operator learns is that the smaller the casting the higher the pouring temperature. A small mold causes relatively rapid cooling of the molten metal and it is essential that the mold cavity is completely filled before the metal starts to solidify. It is impossible to give strict pouring temperature guidelines for all of the many alloys that can be sand cast. The following list will serve for some of the more common metals and alloys.

Approximate pouring temperature (°F.)

Alloy	Small castings	Heavy castings
Yellow brass	1900-2000	1800-1900
Red brass	2100-2200	1900-2100
Tin bronze	2100-2200	1900-2100

Alloy	Small castings	Heavy castings
Manganese bronze	1900-2000	1800-1900
Silicon bronze	2100-2200	1900-2100
Bronwite	1650-1850	not recommended
Aluminum alloys	Not over 1400	Not over 1400

Metals should never be overheated. Watch the temperature closely and remove crucible from furnace, or shut off the furnace as soon as the proper pouring temperature is reached. Skim the metal in the crucible and pour the mold immediately. You can make a skimmer by heating a piece of 1/2-inch diameter steel rod in the furnace and hammering it into a spoon shape.

DO'S AND DON'TS OF POURING

Pour the metal rapidly without excessive turbulence. Try to keep the sprue full to prevent excessive oxidation and dross formation.

Copper-based alloys tend to absorb gases, particularly hydrogen, in gas-fired furnaces. When the metal solidifies it tries to release the gas it has dissolved. This reaction often leads to gas porosity in castings. This effect can be minimized by melting the metal as rapidly as possible and pouring it into the mold as soon as possible. Stirring molten metal will accentuate the gas pickup problem. The furnace should be operated with an excess of air if possible. This condition occurs when a blue or green flame issues from the exit port. A bright yellow flame indicates that the furnace is probably operating with an excess of fuel which may cause excessive hydrogen pickup by the melt. Dirty or oily scrap metal will cause gas pickup problems. Charge only clean, oil-free metal to the crucible.

Remember the following rules for handling and pouring copper-based alloys for best results:

1. Do not agitate or stir the melt.
2. Bring the lip of the crucible as close to the sprue opening as possible during pouring to prevent excessive turbulence.
3. Control the stream of metal to avoid splashing.
4. The metal should be poured in a steady stream without interruption.
5. Keep the sprue completely filled. Reduce the size of the gate if necessary to "choke" the flow and keep the sprue full.

Chapter 7

Core Making

The sand molding operations described so far have been for simple shapes without significant deep recesses or hollow portions. When the part to be made has such configurations, sand cores must be provided in the mold. Core design and production, and the attendant molding operations call forth the highest skills available in the foundry arts. A good example is the production of internal combustion engine cylinder blocks and cylinder heads. The complicated coring involved in the production of these parts almost defies comprehension, yet the auto industry turns out millions of engines each year at remarkably low cost. Credit for the American automobile industry's rapid development in the early part of this century belongs in part to the skilled and dedicated foundry engineers and workers.

Production castings usually call for so-called dry-sand cores. They are made of sand mixed with a binder such as linseed oil, and baked in ovens until rigid enough to handle in subsequent production molding operations. Cores are often formed by packing the sand and binder mixture into wood or metal molds. Complicated cores made in two or more parts stuck together with core paste. After the sand mold has been made in a separate operation, the core, or cores, are assembled in it.

Figure 7-1 depicts a hypothetical casting operation with a dry sand core. Provisions must be made for locking the core into the mold so it cannot shift when the metal is poured. This is done by projections on the pattern which leave so-called "core prints" in

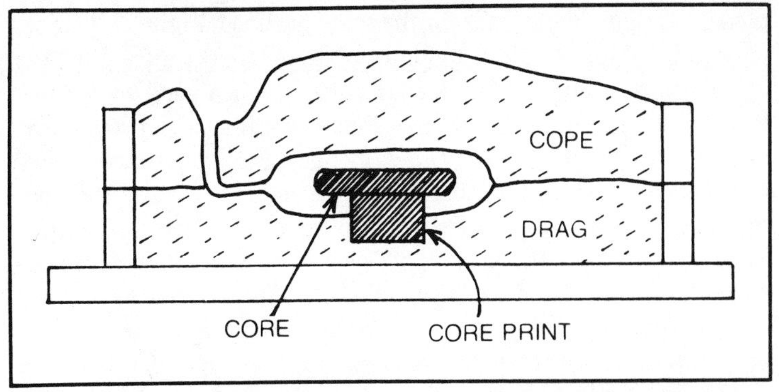

Fig. 7-1. The core is held firmly in position with the core print.

the mold. A corresponding projection on the core fits snugly into the core print, and holds the core firmly in position.

The decision of whether to provide a core or machine out the hollow portion of a solid casting depends on the relative cost of core making and machining. A small hole such as one used for a bolt, stud, or cap screw, usually will not call for the use of a core—the hole will be drilled in the casting after it is made.

SUPPORTING AND MAKING CORES

Cores often require more support than the core print affords. A thin, fragile core needs support against the downward moment that its weight causes and against the upward moment caused by the core's tendency to float on the molten metal when the mold is poured. Thin, hat-shaped metal pieces, called chaplets, usually made of brass, prevent the core from sagging. You can use core

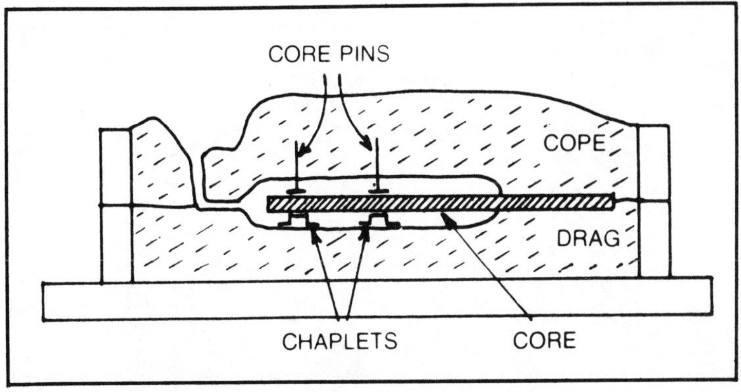

Fig. 7-2. A fragile core supported by chaplets and core pins.

pins resembling brass nails for the same purpose. Figure 7-2 shows how to combine chaplets and core pins to support a fragile core.

To make core sand, mix fifty parts of silica sand by volume to one part linseed oil. You may have to adjust the proportions slightly, depending on the sand's fineness. Foundry suppliers sell core sand, oil, and paste that is used to stick core parts together to make more complicated shapes. When it is mixed thoroughly, the core sand is pressed into properly shaped molds to make the cores. The wet core must be handled carefully to avoid breakage or distortion until it is baked. Cores are baked on a pan or sheet at 300-400 °F. for about an hour, or until hard and dry.

Cores may be formed in plaster, wood, or metal molds. For example, a simple cylindrical core is formed by pressing core sand into a metal tube that has been split lengthwise and reassembled. The tube is taken apart to remove the molded core. Core molds should be liberally dusted with graphite to prevent the wet core sand from adhering to it. More complicated cores can be made in plaster molds.

GREEN SAND CORE MAKING

Relatively simple cores can be made by the "green sand" method. The same molding sand used to make the molds is used to make the cores. No baking is required using this method. These procedures can best be illustrated by showing the actual molding of a radiator cap. Figure 7-3 shows the cross-section of a radiator cap pattern that has been machined from a solid bar of aluminum. The recess in the cap is too deep for the pattern to be withdrawn from the sand mold without breaking the sand; a green sand core will be made using the pattern itself for a mold.

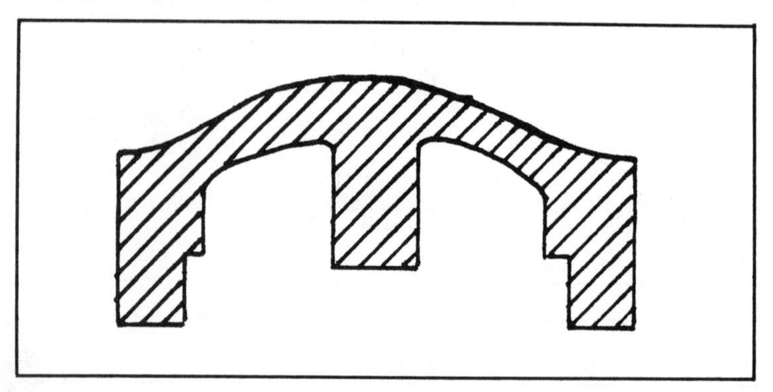

Fig. 7-3. Cross section of a pattern for a radiator cap casting.

Fig. 7-4. The radiator cap casting pattern machined from solid aluminum stock.

Figure 7-4 shows the radiator cap pattern. The cavity is packed firmly with molding sand (Petro Bond sand). A spatula is used to strike off the molding sand so it is even with the pattern cavity's edge. In Fig. 7-5, the pattern is rapped gently in two or three directions to slightly loosen the sand. Then the pattern is inverted and the core allowed to drop out. Dusting the pattern cavity with some graphite powder before packing in the sand makes it easier to separate and remove the core (Fig. 7-6).

The radiator cap pattern's cavity is repacked with sand and

Fig. 7-5. The cavity of the pattern has been firmly packed with Petro Bond molding sand. After rapping the pattern gently against a solid object, the sand is loosened enough for the core to drop out intact.

Fig. 7-6. The green sand core and the pattern.

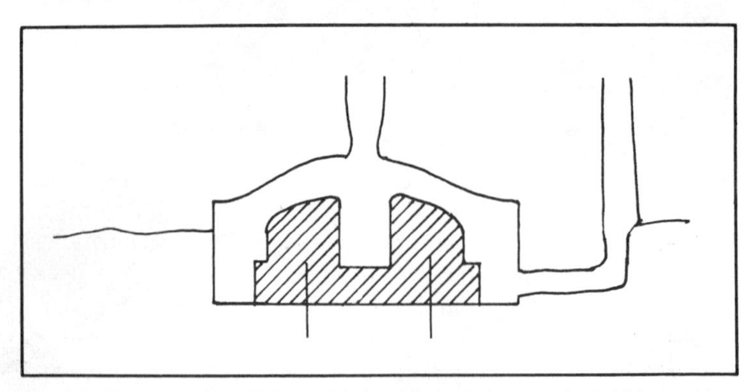

Fig. 7-7. Cross section of the mold showing how the core is held in position with core pins.

Fig. 7-8. Green sand cores and patterns for a gear box housing casting.

used for making a mold the regular way. The green sand core is placed in the mold cavity as shown in Fig. 7-7, carefully centered and held firmly in place by two or three core pins. For simplicity, no core print is used with this pattern, and none is needed because the core pins will sufficiently hold the core.

Figure 7-8 shows examples of more complicated green sand cores with the cores and patterns for the two halves of a gear box casting. Cores were made by ramming Petro Bond sand into the separate halves of the original gear box. Surplus sand in each pattern half was struck off even with the patterns' top edges using a knife blade. The pattern was rapped sharply to loosen the sand and jar the cores out of the respective cavities. The parts were rammed full of sand again, struck off level, and used as patterns to make the molds. After the molds were opened and the patterns removed, the cores were carefully positioned in the cavities of the mold and held in place with core pins as illustrated in Fig. 7-7.

Chapter 8

Casting Problems and Their Causes

Most foundry operations require highly skilled workmanship, so the beginner should not expect perfect results every time. There is no way that you can learn the techniques of the craft by reading a book any more than you could become a skilled mechanic, sculptor, or piano player by reading one. The operator who practices the operations, analyzes the defects in his products, and improves his techniques will soon be turning out professional-quality work. Beginners should start with simple castings and progress gradually to more complicated parts. Starting with complicated castings often leads to discouragement and defeat.

Sand casting is a fascinating activity for the creative person largely because it offers almost limitless opportunities to develop skill and inventiveness. Every new part you make will test your skills as new problems are introduced. Problems will arise even after you have considerable skill. It behooves you to recognize the causes of the problems and take appropriate action to correct them. The following outline provides a guide for some of the most common casting problems. It is applicable primarily to copper-based alloys.

PROBLEM	MOST COMMON CAUSE In order of probability
Misrun: An incompletely filled mold cavity. Corners	1. Metal too cold, interrupted pour, or pouring too slowly

PROBLEM	MOST COMMON CAUSE In order of probability
not filled out or holes through thin sections of casting	2. Inadequate venting 3. Insufficient head of metal 4. Sand too wet or impermeable 5. Gate plugged or too small
Excessive shrinkage, cracks or cavities	1. Metal too hot when poured 2. Inadequate feeding of heavy sections; provide more gates 3. Heavy and thin sections too close without adequate fillets; redesign pattern
Cold shut: poor joining where two streams of metal meet, causing a crack or weakness in the casting	1. Metal too cold when poured 2. Inadequate gating or plugging of gates or runners with dross or oxides 3. Insufficient head of metal 4. Sand too wet or too impermeable 5. Insufficient venting
Sand wash: rough casting surfaces with rough holes at other points caused by sand movement during pouring	1. Too light ramming 2. Weak sand areas around gate or on sharp edges of mold 3. Poor sand 4. Poorly bonded core
Gas holes: large cavities under casting surface	1. Too much gas absorption during melting 2. Too high temperature when poured 3. Metal contaminated 4. Sand too wet
Poor structure: weak castings, badly discolored, or granular fractures	1. Too much gas absorption during melting 2. Inferior quality or contaminated alloy 3. Excessive pouring temperature

PROBLEM	MOST COMMON CAUSE In order of probability
Burning into sand: rough casting surface often with imbedded sand	1. Sand too coarse or permeable, or dry 2. Excessive pouring temperature
Core blow: large gas cavity above a core in casting	1. Insufficient baking of dry sand core 2. Inadequate core venting

Chapter 9

Finishing Castings and Correcting Defects

Almost without exception, every casting requires one or more subsequent operations before it becomes a functional or decorative part. These finishing operations usually are more time consuming and expensive than the actual production of the rough casting.

The first operation after the casting has cooled sufficiently to be removed from its mold is to rid the casting of sand. The easiest way to this with small castings is to brush them under a stream of flowing water. Next, separate the casting from its sprue by clamping the sprue in a vise and cutting it off close to the body of the casting. Remove any risers.

The next operation is to remove or grind off the "flashings," the thin, knife-edged protuberances attached to the rough casting at the mold joint where the cope and drag meet. Any irregularities in the casting caused by minor sand breaks or other imperfections in the mold should be smoothed or ground at this time.

USE A RESINOID GRINDING WHEEL

An ordinary silicon carbide grinding wheel does not work well on brass, bronze, or aluminum because it tends to "load up" with these materials. A resinoid wheel made of layers of abrasive-filled cotton fabric bonded with phenolic or similar bonding agents is better for these operations. The slight resiliency and open structure of the resinoid wheel provides fast cutting action with almost no loading in normal use with brass, bronze, and aluminum. A 9-inch

Fig. 9-1. Grinding flashings off a rough casting with a resin-bonded grinding wheel. The operator must wear a face shield and goggles while performing grinding, polishing, or buffing operations.

wheel, 1/4 inch thick, driven at about 3000 rpm with a one-third horsepower electric motor will handle most work of this kind in the home craftsman's workshop. A grit of 16-36 can be used, but the selection will depend on how smooth a finish is desired.

From now on the work becomes very divergent depending upon the final application of the part. Most parts will require some sort of fastening hole or stud calling for the conventional use of drills, taps, or threading tools. A lot of work can be done with hand tools, but power tools often are essential.

Fig. 9-2. An array of polishing, grinding, and buffing wheels is useful for finishing operations.

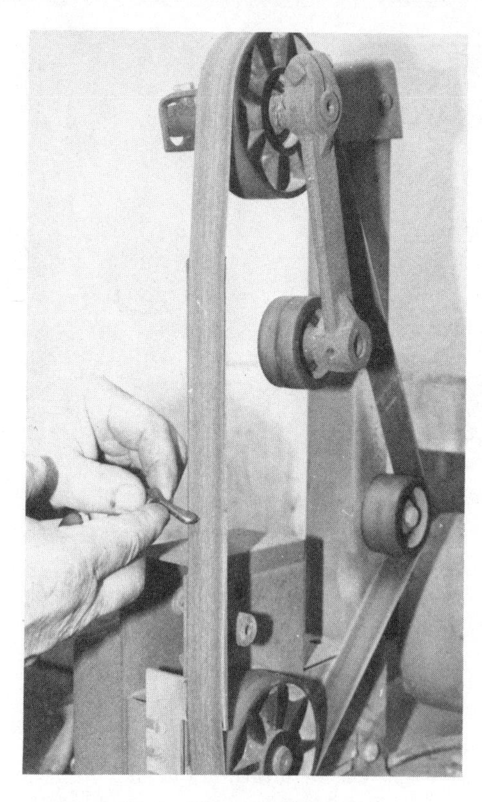

Fig. 9-3. An abrasive belt sander is a versatile tool for finishing small parts.

POLISHING AND BUFFING

Non-decorative castings or ones that need painting, can be left with the as-cast finish and require no further preparation. The majority of reproduction parts castings will require plating, however. The user can let an electroplater polish or buff the rough castings, or he can do the work himself and save the extra cost. Polishing and buffing are mostly hand operations, and the labor costs are high.

The abrasive belt sander shown in Fig. 9-3 is a versatile tool for finishing operations on small castings. Abrasive-charged polishing wheels shown in Fig. 9-4, usually referred to in the metal finishing trade as "set-up" wheels, are capable of producing very smooth finishes on metal parts. These flexible cloth wheels remove metal from metal surfaces without gouging them the way a rigid grinding wheel will. The floppy Grind-o-Flex sandpaper wheels will perform much the same way as the set-up wheels.

For finely detailed small casting finishing, the operator will find that the Dremel, Foredom, or Dumore hand grinders (Fig. 9-5) are indispensable. You can fit these grinders with a wide variety of

Fig. 9-4. Polishing a part on a cloth wheel. This type of polishing wheel with abrasive grain (alumina) bonded to the periphery with glue produces smooth finish on irregularly shaped parts.

burrs, sanding discs, and flexible polishing wheels. The tools are capable of finishing small details, depressions, or cavities with extreme delicacy that is impossible with any other equipment.

You should seek other sources of information for complete details about polishing and buffing castings to prepare them for plating. The information is beyond the scope of this book, but manufacturers of polishing and buffing supplies and equipment usually have bulletins detailing the use of their products.

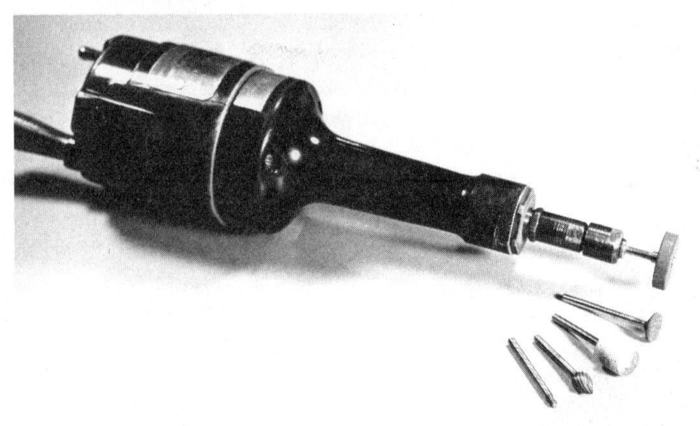

Fig. 9-5. A Dumore hand grinder is useful for detailed finishing of small parts.

CORRECTING DEFECTS IN CASTINGS

Castings will sometimes have minor defects such as sand pits, gas holes, and shrinkage cracks or cavities that cannot be polished out without removing too much metal. Sometimes the defects are not apparent in the rough casting but show up when the part is polished. The presence of severe defects is reason for rejecting the part, but soldering or brazing can correct a minor imperfection. Salvaging castings by welding and brazing is a common practice in industry, so you need not feel embarrassed by doing a little repair work on your castings. If the part is to be painted, you can ignore the minor defects after filling them with spot putty or body filler resin. This kind of repair is not suitable for parts to be plated because it interferes with the plating process.

Before the holes or pits can be filled with solder or braze, they must be drilled out with a small twist drill or a dental burr in a hand grinder. This provides sound metal in the cavity for the solder or brazing alloy to adhere to. Soft solder (lead-tin) may be used to fill holes in brass or bronze castings, but because it is softer than the surrounding metal the solder has a tendency to polish out and may leave a perceptible indication of the repair after plating. It is better to use silver brazing alloy to fill defects because the silver braze is more nearly the hardness of the surrounding metal and will not polish out of the repair.

Figure 9-6 shows a repair on a sand cavity left in a robe rail

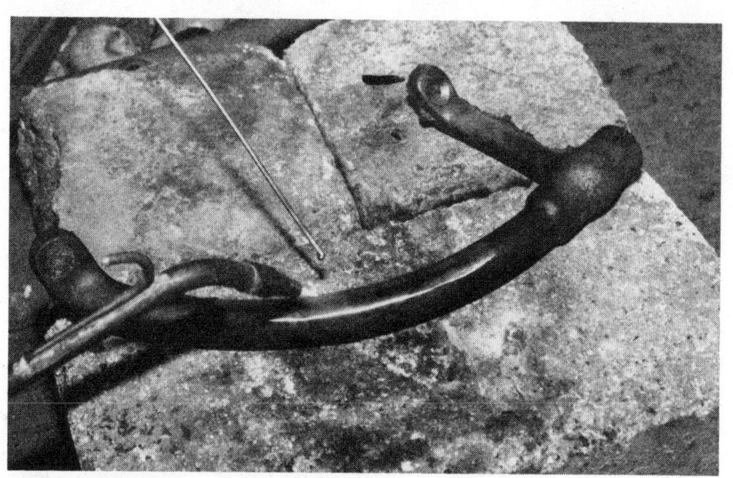

Fig. 9-6. Repairing a sand cavity in a defective casting by silver brazing. The cavity must first be ground or drilled out to provide sound base metal for the braze to adhere to.

bracket casting. If the sand cavity is properly filled with silver brazing alloy and the defect is polished down to the level of the surrounding metal, the repair should be invisible after plating.

MAKING MORE INTRICATE PARTS BY BRAZING

An earlier chapter emphasized that the complexity of a part to be produced by sand casting is limited by the requirement that the pattern have a so-called "line of parting." This way the mold's two halves can be separated without breaking the sand in the mold. This requirement usually is met by parts that were originally produced by sand casting, so reproducing them ordinarily presents no problem.

Parts originally made by other casting methods, die casting in particular, may not have a line of parting. The casting pattern made with these parts usually cannot be withdrawn from the two halves of the mold without disturbing the sand. These parts ordinarily are reproduced by lost wax (investment) casting or by casting in multiple-piece plaster molds. The alert technician need only be aware that it is possible to section the pattern into two or more parts that *can* be made by sand casting. When they are made, the parts are joined by silver brazing to produce the final reproduction part.

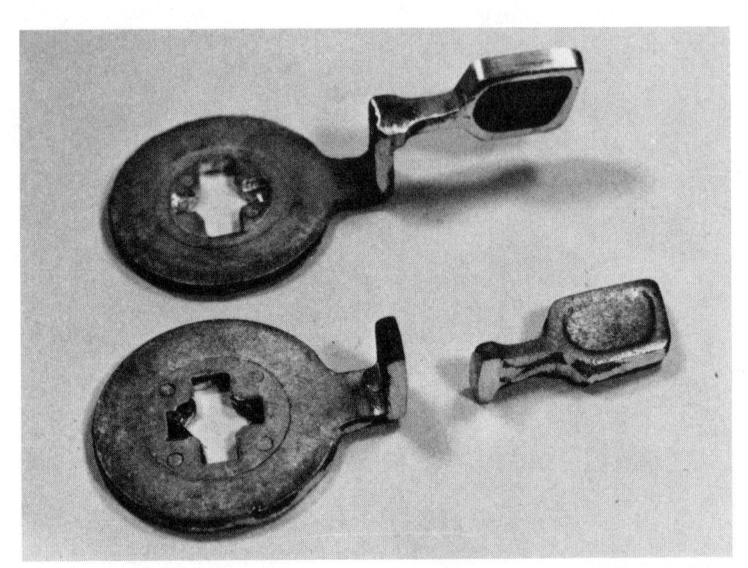

Fig. 9-7. This light switch handle cannot be readily cast in one piece by sand molding because there is no parting line. Two separate sand castings are made (below) and joined by silver brazing to produce the finished part (above).

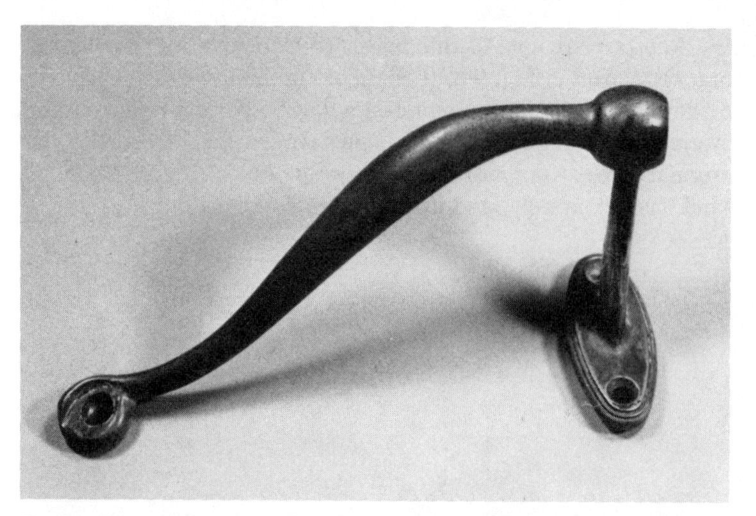

Fig. 9-8. This robe rail bracket cannot be sand cast in one piece because there is no line of parting. It was originally made by die casting in zinc alloy.

If properly made, silver-brazed joints in bronze or brass are as strong as the base metal, so the part's integrity is as good as if it were made in one piece by the more labor-intensive lost wax casting method.

A good example of this technique is shown in Fig. 9-7 showing production of a light switch handle from an antique car. The plate and handle are cast separately in sand molds. The handle is attached to the plate by a mortised silver brazed joint. This method is far less expensive and time consuming than investment casting for production of a few pieces.

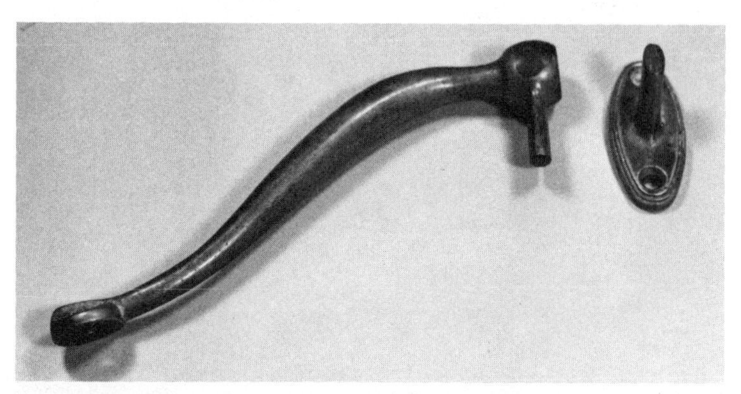

Fig. 9-9. By sectioning the original part to produce two patterns which can be sand cast, the original part is reproduced by joining the parts by silver brazing.

Another example of this powerful technique is the robe rail bracket shown in Fig. 9-8. It is almost prohibitively expensive to make this part by investment casting, but a perfect replica of the original part can be made quickly and economically by cutting the original part pattern into two pieces (Fig. 9-9) which can easily be sand cast separately, and joining them by silver brazing.

Chapter 10

Do's and Don'ts for Safety

It should be obvious that foundry operations are potentially hazardous. Molten metal at a temperature near 2000 °F. will instantly ignite any combustible material that it contacts—paper, wood, cloth, plastics. You must keep all combustible material of this sort away from the area where melting and pouring takes place. It is a good idea to have some dry sand ready to sprinkle on the wooden parts of the flask to extinguish any small fires that may occur.

Pay utmost attention to avoiding any personal injury. The operator who is performing melting and pouring operations must wear goggles and faceshield, heavy leather shoes, leather or asbestos gloves, and flame-resistant clothing.

The following precautions are also essential:

1. Do not use damaged or cracked crucibles that may break and release molten metal when they are lifted.

2. Do not allow moisture to come into contact with molten metal. Do not plunge any damp objects into molten metal. To do so may cause a steam explosion that can spew molten metal. Keep water away from the melting area, and do not use it to extinguish small fires. Rely on dry sand instead.

3. Do not add cold chunks of metal to the molten metal because condensation of burning gases may cause moisture to form on the surface of the cold metal. Warm cold metal slowly by placing it on top of the furnace for a short time.

4. When pouring castings, arrange some dry sand around the

flask. In case molten metal escapes it will be stopped without doing any damage.

5. Guard against gas leakage or accumulation of unburned gases and carbon monoxide. Avoid operating in an enclosed or unventilated area.

6. Do not pour molten metal onto a wet or damp surface, other than molding sand which has a minimum amount of conditioning moisture necessary for adequate cohesion.

7. Do not melt or vaporize any metals or chemical compounds with uncertain or unknown composition. It may generate toxic gases or fumes. Beryllium, cadmium, or lead may generate highly poisonous fumes. Zinc fumes ordinarily are not considered highly toxic, but it is good to avoid inhaling them because they have an adverse physiological effect on some individuals.

8. Do not charge magnesium to the furnace without full information about proper melting precautions. As far as the beginner is concerned, it is best to avoid melting magnesium or magnesium alloys altogether.

9. Wear goggles when you saw, grind, polish, or buff finished castings.

Chapter 11

Foundry Terms

abrasive: A natural or artificial substance used for grinding, polishing, buffing, lapping, or sandblasting. Commonly includes garnet, emery, corundum, diamond, aluminum oxide, and silicon carbide.

aging: In a metal or alloy, a change in properties that occurs slowly at room temperature and more rapidly at higher temperatures.

alloy: A substance with metallic properties, composed of one or more chemical elements, at least one of which is a metallic element.

alumel: A nickel-based alloy frequently used as a component of thermocouples.

annealing: Heating and holding at a suitable temperature and then cooling at a suitable rate, usually to reduce hardness, to improve machinability, or to achieve other desired properties.

baking: Heating at a low temperature to remove gases.

base metal: The metal present in the highest proportion in an alloy. Brass, for example, is a copper-base alloy.

belt grinding: Grinding with an abrasive belt.

bentonite: A clay-like substance used as an ingredient in molding sands.

binder: A material, other than water, added to molding sand to bind the particles together.

blasting: Cleaning or finishing metal by impingement with abrasive particles carried by gas or liquid.

blind riser: A riser that does not extend through the top of the mold.

blowhole: A hole in a casting caused by gas trapped during solidification.

bottom board: A flat board used to hold the flask when making molds (usually called molding board).

brass: An alloy consisting mainly of copper (over 50 percent) and zinc, to which smaller amounts of other metals may be added.

brazing: Joining parts by flowing a thin layer of non-ferrous filler metal in the space between them. The term brazing is ordinarily used if the process is carried out above 800 °F.; below this temperature it is called soldering.

Brinell hardness test: A test for a material's hardness by forcing a hardened steel or carbide ball of specified diameter in to the material under a specified load.

bronze: A copper-based, tin alloy with or without other elements. Some tinless alloys are sometimes referred to as bronzes. The term is loosely applied.

buffing: Developing a lustrous surface appearance by contacting the metal with a rotating buffing wheel.

buffing wheel: Fabric, leather, or paper discs held together, usually by sewing, to form wheels for grinding, polishing, or buffing.

bumping: Ramming sand into a mold by jarring or jolting.

burnt-in sand: A casting defect caused by sand adhering to the casting's surface.

capillary attraction: A combination of forces which causes molten metals or other liquids to flow between closely spaced solid surfaces.

carbide: A carbon compound with one or more metallic elements.

casting: (noun) An object obtained by solidification of a substance in a mold. (verb) Pouring a substance into a mold to obtain an object of the desired shape.

casting shrinkage: The reduction in volume of a metal as it solidifies and cools.

casting stresses: Stresses set up in a casting primarily caused by shrinkage.

catalyst: A substance that changes the rate of a reaction without itself undergoing any net change.

centrifugal casting: A casting made by pouring metal into a rotating mold.

chaplet: A metal support for holding cores in place in sand molds.

cheek: An intermediate section of a flask used between the cope and the drag when molding a shape that requires more than one parting line.

chill: A metal insert placed in a sand mold to increase the cooling rate at that point.

chromel: A nickel-chromium alloy used for thermocouples and heating elements.

clay: An earth substance consisting mainly of hydrous aluminum silicate and often used as a binder in molding sands.

cold shut: A discontinuity in the surface of a casting caused by two streams of molten metal failing to unite.

cope: The upper or topmost section of a flask, mold, or pattern.

core: A formed section inserted inside a mold to shape the interior of a casting.

core blower: A machine for making foundry cores.

Croning process: A shell molding process.

crucible: A pot or vessel used for melting metal or other substances.

crush: A casting defect caused by partial displacement of sand in a mold before the molten metal is poured.

defect: A condition that impairs the object's usefulness.

degasser: (or degasifier) A material added to molten metal to remove dissolved gases that otherwise might be trapped when the metal solidifies.

degassing: The act of removing dissolved gases from molten metals.

dendrite: A crystal with a branching, tree-like pattern often seen in slowly cooled castings.

deoxidizer: A substance added to molten metal to remove dissolved oxygen.

die casting: A casting process that uses pressure to force molten metal into the cavity of a metal mold.

draft: Taper on surfaces of a pattern that allows it to be withdrawn from the mold.

drag: The bottom section of a mold, flask, or pattern.

drop: A casting defect caused by sand dropping from the cope.

dross: The scum that forms on the surface of a molten metal due to oxidation or when impurities rise to the surface.

dry sand mold: A mold made of sand and then dried.

ductility: The ability of a material to deform without fracturing.

dusting: Applying a powder such as graphite to a mold surface.

elasticity: The property of a material allowing it to regain its original shape after deformation.

emery: An impure form of aluminum oxide used as an abrasive.

erosion: A casting defect caused by the scouring action of flowing metal.

facing: Special sand placed in direct contact with the pattern to improve the surface finish of a casting.

fines: Sand grains substantially smaller than the predominate size in a sand mixture.

finish: A metal's surface condition, quality, or appearance.

flask: A metal or wood frame used to make a sand mold.

fluidity: The ability of a metal to flow into and fill a mold.

flux: A material used to remove undesirable impurities from a molten metal, or a protective cover for molten metals.

foundry: A place where castings are made.

gagger: A metal piece used to reinforce or support the sand in a mold.

gas pocket: A cavity in a casting caused by trapped gas.

gassing: Evolution of gasses from a metal during solidification.

gate: The portion of the runner where molten metal enters the mold cavity.

gated pattern: A pattern which includes a mold's gate.

grain refiner: A substance added to molten metal to attain a finer grain structure in the casting.

grinding: Removing stock from work using a grinding wheel.

grinding wheel: A circular cutting tool made of bonded abrasive grains.

grit size: The nominal size of abrasive particles determined by the number of openings per lineal inch in a screen through which the particles will pass.

holding furnace: A small furnace in which molten metal is transferred and held until ready to pour.

hot tear: A fracture formed in a casting during solidification because the casting is restrained from shrinking.

ingate: Same as gate.
ingot: A casting used for remelting.
insert: A removable portion of a mold.
investment casting: Casting metal into a mold that is made by surrounding (investing) an expendable pattern (usually wax) with a refractory slurry which sets, after which the pattern is melted out. Also called "lost wax" casting.
investment compound: A mixture of refractory filler, binder, and liquid used to make molds for investment casting.

jacket: A wood or metal form slipped over a sand mold for support, especially during pouring.

ladle: A receptacle for transferring or pouring metal.
loam: A molding material made of sand, silt, and clay used for making very large castings.
lost wax process: Investment casting in which a wax pattern is used.

match plate: A plate of metal or other material upon which patterns are mounted to facilitate molding operations.
melting point: The temperature at which a pure metal, compound, or eutectic changes from a solid to a liquid.
metallurgy: The science and technology of metals.
misrun: A defective, partially formed casting caused by solidification of the metal before the mold cavity is filled.
mold: A form made of sand, metal, or other material with a cavity into which molten metal is poured to form a casting.
molding machine: A machine used to make molds by mechanically compacting sand around a pattern.
mold wash: An emulsion of various materials used to coat a mold cavity's surfaces.
mulling: Mixing sand and clay by a rubbing or rolling action.

neutral flame: A gas flame in which there is neither an excess of fuel nor air.

oxidizing flame: A flame with an excess of air (or oxygen).

parting dust: A composition used to ease the separation of the pattern in sand molding and to prevent the sand from sticking at the junction of the cope and drag.

parting line: A plane on a pattern corresponding to the separation between the cope and drag.

pattern: A form of wood, metal, or other material around which molding material is placed to make a mold.

permanent mold: A metal mold that is used repeatedly for the production of castings.

pickling: Removing surface oxides from metals by chemical action.

pig: An ingot.

pipe: A central cavity formed in a casting during solidification.

plaster molding: A process for making a mold by forming a slurry of gypsum (Plaster of Paris) around a pattern, allowing it to harden and to dry thoroughly.

plumbago: A high quality graphite powder.

porosity: In a metal, fine holes or pores; in a sand, degree of permeability to gases.

pouring: Transferring molten metal from a ladle or crucible to a mold.

pouring basin: A basin or funnel located on top of a mold to receive molten metal.

precision casting: A metal casting of accurate, reproducible dimensions.

precoat: A special refractory coating applied to wax patterns in investment casting.

pressure casting: Making castings by applying pressure to the molten metal as in die casting.

pyrometer: A device for measuring temperatures above the range of thermometers.

quenching: Rapid cooling.

rabbling: Stirring molten metal with a tool.

ramming: Packing sand into a compact mass.

reducing flame: A gas flame produced with excess fuel.

refractory: A material with high melting point suitable for use in molds and furnace linings.

resinoid wheel: A grinding wheel bonded with synthetic resins.

riddle: A sand sieve used in a foundry.

riser: A reservoir of molten metal attached to a casting to provide the additional metal required because of shrinkage during solidification.

runner: A channel for molten metal, usually the portion of a mold connecting the sprue with the gate.
runout: The accidental escape of molten metal from a mold.

sag: A casting defect caused by insufficient sand strength.
sand: A granular material formed when rocks disintegrate. Foundry sands are mostly pure silicon dioxide. Molding sands contain clay.
scab: A casting defect that occurs when a thin layer of metal separates from the casting.
scrap: Discarded metal that can be reclaimed by melting.
sea coal: Finely divided coal sometimes added to molding sands.
semipermanent mold: A metal mold that uses sand cores.
shakeout: To remove castings from sand molds.
shell molding: Croning process. To bring thermosetting resin-bonded sand mixtures into contact with a hot pattern to form molds.
shift: Casting defect caused by mismatching the cope and drag.
shot: Small spherical pieces of metal.
shrinkage cavity: A void in a casting caused by shrinkage.
shrinkage cracks: Hot tears in a casting caused by shrinkage.
shrinkage rule: A measuring rule with expanded graduations to compensate for shrinkage of a casting as it cools.
silver brazing: Brazing with silver-based alloys.
shrinkhead: Same as riser.
skim gate: A gate designed to prevent slag from passing into the mold.
skimmer: A spoon-shaped tool for removing dross from the surface of a molten metal.
skull: The solidified metal or dross left on the walls of a crucible when the molten metal is poured out.
slag: Dross.
slag inclusion: Slag or dross trapped in a solidified casting.
slush casting: A hollow casting, usually made of low melting metal. When the desired thickness of metal has solidified on the walls of the mold, the balance of the molten metal is poured out.
snagging: Free-hand grinding to remove flashings from castings.
snap flask: A flask hinged at one corner so that it can be quickly separated from the mold.
solidification shrinkage: The decrease in metal's volume when it solidifies.
sprue: The channel that connects the pouring basin with its run-

ner. Sometimes the definition includes all gates, risers, and runners.

stress raiser: (or stress riser) Changes in a part's contour that cause localized stress.

tarnish: Surface discoloration of a metal caused by an oxide formation.

tensile strength: The ratio of the maximum load a bar of metal can withstand to the original cross-sectional area.

thermocouple: A temperature measuring device made of two dissimilar metals that produce a voltage or current roughly proportional to the differences in temperature of the hot and cold ends.

tolerance: The permissible variation in a part's size.

vent: A small opening in a mold for the escape of gases.

wash: A coating sometimes applied to a mold cavity.

wildness: A condition that occurs when molten metal releases so much gas that it becomes violently agitated.

Part 2

Casting Rubber Parts

Chapter 12

Rubber and a Space-Age Substitute

Unlike the art of metal casting, which pre-dates recorded history, rubber technology is largely a modern development, mostly of the 19th and 20th centuries. Over 400 years ago Spanish explorers in the New World observed that South American natives played games with an elastic ball made of liquid that oozed from certain trees. The material was then completely unknown in Europe.

In 1731, a French official in the Amazon region sent a quantity of a resinous material collected from the native *Hevea* tree to France. He reported that Amazon natives used the material called "caoutchouc" to waterproof fabrics and to make footwear. He also reported that they produced watertight bottles by covering clay molds with the material. When the resin dried they smashed the molds and removed the pieces through the bottle neck. This example must be one of the first recorded methods of casting rubber parts.

Some of the crude resin eventually reached England. When the chemist Joseph Priestley examined it and found that it was capable of rubbing out pencil marks, he coined the name "rubber." In the early 1800s, a small rubber clothing and footwear industry developed, but the apparel deteriorated rapidly, and was so stiff in winter and gummy in summer, that consumers were disgusted with the products.

DISCOVERY OF VULCANIZATION

It was not until 1839, when Philadelphian Charles Goodyear

discovered the vulcanization of rubber that rubber began to enjoy a measure of commercial acceptance. Goodyear's discovery, considered one of the most important technological inventions of the 19th century, showed that when rubber is heated in the presence of sulfur, it is highly stable in the heat and in the cold. Within a few decades, the rubber industry expanded to produce thousands of useful consumer items.

In the early days of the rubber industry, most of the raw material came from Brazilian jungle trees. The uncertain supply, variable quality, and price fluctuations of Brazilian rubber spawned British and Dutch rubber plantations in Ceylon and Sumatra that gradually replaced wild rubber production.

Natural rubber is a material that cannot be readily duplicated in the laboratory. Nature alone produces it in the cells of certain plants. Nevertheless, as early as 1890 chemists were experimenting with methods for producing synthetic rubber. Results were discouraging at first because of the high costs, and it was not until 1931 that the duPont Company produced Neoprene, the first commercially successful synthetic rubber. Neoprene closely resembles natural rubber, but it is more resistant to the action of ozone and oil. It is still used today.

The loss of natural rubber sources during World War II provided a tremendous impetus for the invention and development of many new synthetic rubbers. Many of the synthetic rubbers possess properties far superior to natural rubber, and often are used for special applications where the natural product is unsuitable.

PRODUCTION OF RUBBER PARTS

Most of the molded rubber parts of commerce are made by mixing raw rubber stock—whether natural, synthetic, or reclaimed—with appropriate fillers, extenders, and vulcanizing agents. The mixture is forced into steel molds and heated under pressure to vulcanize the rubber. Generally, rubber does not lend itself to economical small-scale production because of costly mixing equipment, molds, and other production machinery. You cannot expect to undertake this activity on a small scale to produce a few parts for your own use.

Rubber inevitably will undergo degradation in normal use, particularly when exposed to heat, sunlight, and atmospheric ozone because it is an organic material. The restorer of automotive vehicles today will find considerably degraded rubber parts on nearly all cars more than ten or twenty years old. The restorer who wishes to return his vehicle to its original pristine condition must either petition rubber parts manufacturers to produce suitable parts, or adapt some form of commercially-available item to his use. A small, but brisk reproduction rubber parts business has developed, but for economic reasons the manufacturer can profitably make only parts in great demand. He has to sell a lot of parts to amortize the cost of molds and other production equipment. Owners of some obscure makes of cars are out of luck if not enough demand exists for the parts they need.

A liquid form of rubber that can be poured into simple molds and cured without heat or pressure is needed so that the amateur can make his own parts. Rubber cannot be prepared in liquid form because of its chemical nature. It can be dissolved in solvents to make a liquid, but the solution is worthless for casting parts because the shrinkage that occurs when the solvent evaporates destroys the parts' shape. Some silicone rubber meets the necessary requirements, and rubber parts can be cast from it. In general, though, silicone rubber lacks the strength, hardness, and abrasion resistance we expect from rubber. As a result, its use is limited to applications that use silicone's tremendous resistance to the effects of high and low temperatures.

THE ANSWER: POLYURETHANE RUBBER

Polyurethanes are versatile polymer materials that are produced in a variety of forms ranging from soft foams to hard plastics, depending on the desired end-use requirements. They are made from isocyanates and organic compounds containing hydroxyl groups. Polyurethanes are almost invariably supplied as two-component systems of resin and curing agent that are mixed immediately before use.

It is a moot question whether it is proper to refer to polyurethane as "rubber." Technically, it has little resemblance to the chemical composition of natural or synthetic rubbers, but when it is properly compounded, polyurethane resembles rubber and has many of its properties. Common usage provides precedence for calling it rubber, and we shall do so in this book.

Polyurethane rubber is available through commercial sources in one- and ten-pound kits under the trade name Flexane, manufactured and distributed by the Devcon Corporation of Danvers, Massachusetts. Each 1 pound Flexane kit (Fig. 12-1) includes separate containers of resin and curing agent, and a plastic mixing container with a flat wooden spatula for mixing (Fig. 12-2).

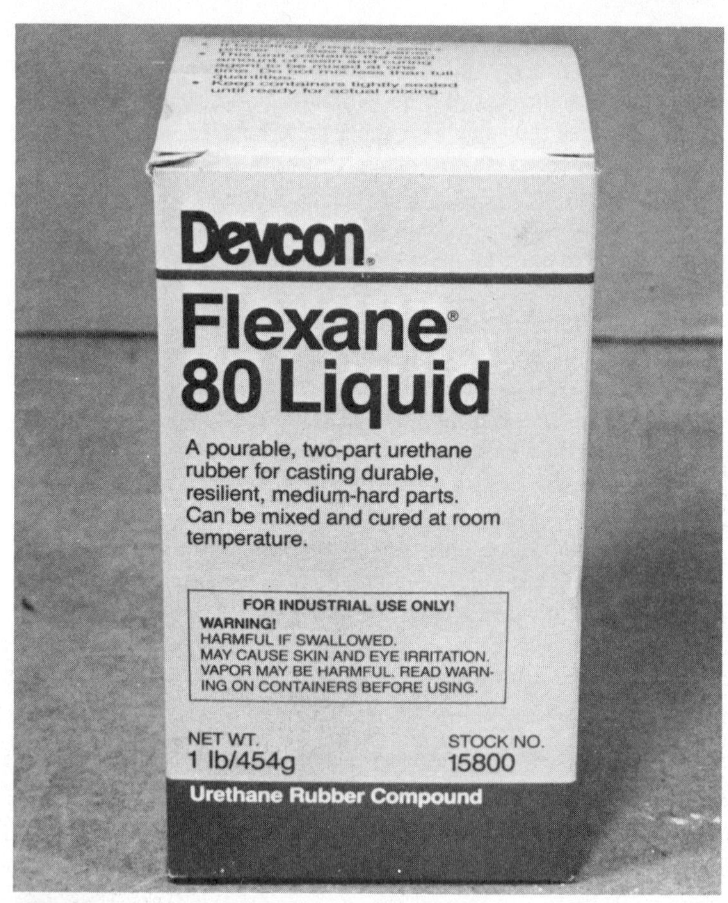

Fig. 12-1. A 1-lb. kit of Flexane 80 Liquid polyurethane rubber.

Fig. 12-2. Contents of the 1-lb. Flexane 80 Liquid polyurethane rubber package.

The Flexane polyurethane rubber is mixed and cured at room temperature and requires no heat, pressure, or special expensive molds. Flexane is supplied in putty and liquid forms. Flexane putty is designed to be puttied or troweled into place, while the liquid can be poured into molds for making small rubber parts. When cured, polyurethane rubbers have exceptionally high strength, and are tear and abrasion resistant. Fully cured, it looks and flexes like conventional natural or synthetic rubber.

THE FLEXANE POLYURETHANES

The availability of Devcon Flexane in handy one-pound kit form opens up new vistas for the home craftsman who reproduces car parts, experimental parts, and repair parts for equipment and tools. The potential applications are limited only by the inventiveness and ingenuity of the amateur technician. In this book, we can do little more than touch on a few typical examples that will illustrate basic applications and principles.

Flexane is made in two hardness grades: 80 and 94. The numbers correspond to Shore A durometer hardness of the cured rubber. The Shore Durometer is an instrument used to measure the hardness of plastics and rubber by pushing a blunt spring-loaded needle into the materials. The distance that the needle penetrates is a measure of the hardness and is indicated by a pointer on the face of the instrument. As an example, a soft rubber gasket or automobile door weatherstrip may have a durometer of about 30. A rubber automobile tire is about 50-60, while the cover of a golf ball is about 95 and can scarcely be indented by the pressure of a fingernail. These are only crude estimates, but give you some idea of the range of rubber hardness. Flexane 80 may be classed as a medium hard rubber when cured. Flexane 94 is a hard rubber and would be restricted for use in such parts as rubber knobs, gear shift knobs, or steering wheel covers in automobile restoration. Most typical reproduction car parts like door bumpers, pads, mats, grommets, gaskets, and weatherstrips call for a durometer hardness of about 60 to 70.

A flexibilizer—Flex-Add—(Fig. 12-3) can be used with Flexane 80 to produce any durometer hardness below 80. Table 12-1 shows how much Flex-Add to mix with each pound of Flexane 80 liquid to obtain the desired hardness.

If a particular reproduction car part calls for a durometer hardness of approximately 60, for example, you would add six ounces of Flex-Add to each pound of Flexane 80 Liquid to obtain a

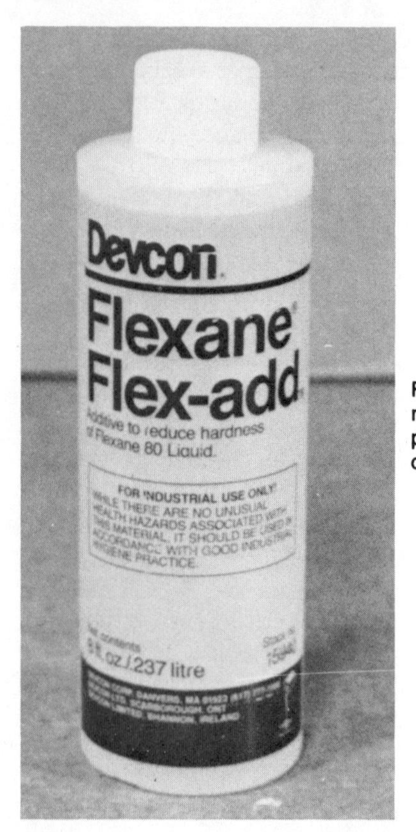

Fig. 12-3. Flex-Add polyurethane may be added to Flexane 80 Liquid polyurethane to produce any desired durometer hardness below 80.

durometer hardness of 61 in the final product. The manufacturer says that the adding of Flex-Add increases the pot life and cure time, but specific numbers are not supplied.

Flexanes do not contain any solvents or diluting agents that must dry so sections of any thickness can be cast simultaneously. Curing produces no significant dimensional change, so you can

Table 12-1. Flex-Add Chart.

Use the following chart to determine the amount of Flex-Add to use with one pound of Flexane 80 Liquid to obtain the desired durometer:

Flex-Add	2 oz.	4 oz.	6 oz.	8 oz.	10 oz.	12 oz.	14 oz.	16 oz.
Durometer	77	69	61	54	49	43	37	30

make a precise copy of the original pattern used for making a mold.

Flexanes adhere strongly to other rubbers, wood, fabrics, glass, and nearly all metals. When casting the material into a mold, use a release agent to prevent the rubber from sticking to the mold. Devcon sells a mold release agent, but silicone grease, Teflon spray, or automobile wax may be used for this purpose. To improve the adhesion of Flexane to metals, for example in making a composite metal-rubber part, Devcon makes Flexane Metal Primer.

Flexanes are waterproof and have excellent resistance to oils, greases, and most chemicals. It has poor resistance to long-term exposure to gasoline, acetone, and chlorinated solvents. The properties of Devcon Flexanes are given in Table 12-2.

MIXING LESS THAN 1-POUND QUANTITIES

A few years ago, Devcon provided mixing instructions with its Flexane kits for the user who wanted to mix less than a one-pound quantity and save the rest for future use. Today, the kits list the following instructions: "This unit contains the exact amount of resin and curing agent to be mixed at one time. Do not use less than full quantities." It behooves the user to plan his work ahead and provide sufficient molds to use substantially all of the package at one time after the resin and curing agent are mixed.

It's not clear why Devcon changed its policy concerning mixing less than unit quantities. The reason probably relates to the limited shelf life of the product once the container of resin is opened. The resin probably is sealed in an inert atmosphere at the factory,

Table 12-2. Physical Properties of Devcon Flexane.

Flexane Type Available As	Flexane 94 Liquid & Putty	Flexane 80 Liquid & Putty
Viscosity with Hardener cps	6,000	10,000
Hardness Shore A ASTM D2240	94	80
Pot Life of 454g (1lb) @ 24°C (75°F) in Minutes	10	30
Demolding Time @ 24°C (75°F) in Hours	3	10
Specific Volume cm^3/kg (in^3/lb)	939 (26)	939 (26)
Cured Density g/cm^3 (lb/in^3) /ASTM 792	1.10 (.040)	1.08 (.039)
Maximum Operating Temperature °C (°F)	121 (250)	121 (250)
Cure Shrinkage ASTM D2566 cm/cm or in/in	0.0004	0.0007
Elongation ASTM D412%	250	350
Compression Set ASTM D395%	7	23
Tensile Stress at 100% Elongation ASTM D412 kgf/cm^2 (psi)	68.25 (970)	33.25 (473)
Tensile Strength ASTM D412 kgf/cm^2 (psi)	105 (1493)	77 (1095)
Tear Resistance ASTM D624 kgf/cm (lb/in)	89.47 (500)	50 (280)
Abrasion[1] Resistance Weight Loss	0.298	0.285
Insulation Resistance ASTM D257 ohms	5.0×10^{13}	4.0×10^{13}
Dielectric Strength volts/mil	340	350

[1]Loss — Mg/1000 Rev. Tabor Abraser 18H Wheel

so that once the container is opened, oxygen from the air accelerates the curing process, rendering the resin unsuited for use after a comparatively short time. Although Devcon warns against the practice, it is possible to mix part of the package's contents and save the rest for later use.

For less-than-full kit contents, mix the resin and curing agent in the ratio of ten parts (by weight) resin to three parts curing agent. Be sure to stir the contents of the curing agent container well before use because the pigment it contains may have settled out in storage. Materials must be weighed as accurately as possible. Use a chemical balance (Fig. 12-4), photographic scale, or a postal scale for weighing the ingredients. Try to use the remainder of the resin within a few days if possible. When you replace the lid on the resin can, do not force it too tightly or it will be hard to get it off a second time because resin in the groove of the can solidifies.

As an example, you can mix 5 ounces of resin with 1-1/2 ounces of curing agent to produce the right proportions. If you want a durometer of less than 80, add the required amount of Flex-Add according to the instructions given in the table above. They are also given on the Flex-Add bottle. To make a durometer of 61, add Flex-Add in the proportion of six ounces to one pound of mixed resin and curing agent. In this case that would be 2.43 ounces. This combination will produce 8.9 ounces of mixture that will yield a rubber of 61 durometer hardness when cured.

Thorough mixing of the resin, curing agent, and Flex-Add (if used) is essential if you want good results. Use a flat-bladed tool like a spatula, and stir well along the sides and bottom of the container. Try to avoid mixing in too much air that will cause bubbles

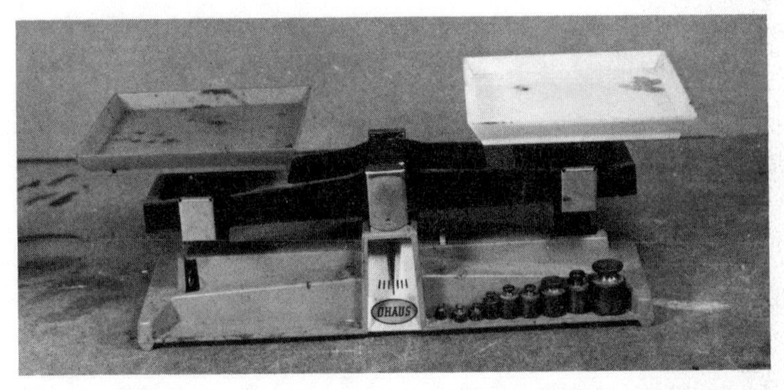

Fig. 12-4. When mixing less than full 1-lb. quantities of Flexane polyurethane rubber, a chemical balance is recommended to accurately measure resin and curing agent.

in the finished part. Soft spots in the cured rubber indicate poor mixing.

WHERE TO BUY FLEXANE

Flexane is an "industrial use only" product, and it is sold mainly through industrial tool supply outlets. It is not regularly sold in hardware or auto supply stores and there apparently is no reliable mail order source for Flexane. Try your phone directory for the name of a local Devcon sales representative, or write or call Devcon Corporation, Danvers, Massachusetts, 01923 (Phone 617-777-1100) for the address of a jobber or sales outlet in your area.

Chapter 13

Making a Top Bow Rest Pad

Making rubber parts out of polyurethane may seem a far cry from casting metal parts in sand molds using molten metal near 2000° F. If we review the definition of a casting (an object produced by solidification of a substance in a mold), you can see that there is no fundamental difference between a metal casting and a rubber casting. Polyurethane rubber casting is technically simpler because producing high temperatures is not required. Sand molds, necessary for refractoriness in metal casting, are never used for rubber casting, but the general principles are the same.

No specific recommendation is given for reproducing commercially-produced rubber parts. It is usually cheaper to buy rubber parts, but this book aims to provide you with valid alternatives in case commercial parts are not available.

A SIMPLE PROJECT

Let's start with a simple project—reproducing the top bow rest pad found on roadsters and convertibles of the twenties and thirties. Two pads are set into a pair of metal brackets bolted to the rear deck. The top bow rests on these pads when the top is lowered, preventing the top bow from touching the vehicle's finish. The top bow rests have a variety of forms and designs depending on the make or model of the vehicle, but they are usually similar to the simple rectangular rubber bumper illustrated in Fig. 13-1. Occasionally, the top rest pad is a flat strip of rubber placed into an in-

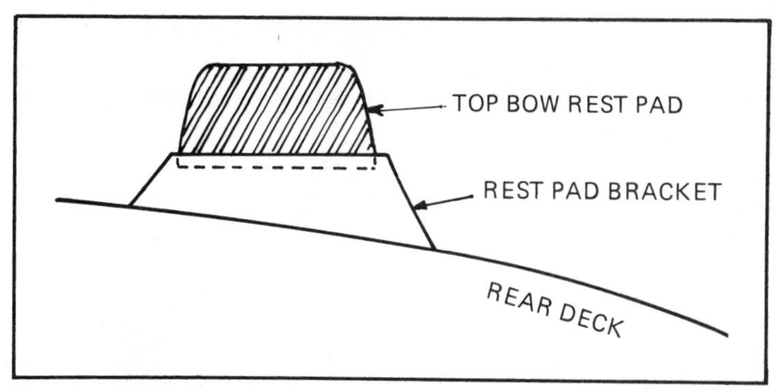

Fig. 13-1. Sketch of typical top bow rest pad and bracket used on rear deck of antique roadster.

set of the hand rails or decorative strips on the rear deck.

The original pad to be reproduced was missing, but the general shape was discernible from photographs. A pattern was made of the part using aluminum that was cut and ground to shape and polished. A photograph of the finished pattern is shown in Fig. 13-2. A hole was drilled and tapped in the center of the base of the pattern for a 10-32 machine screw that serves as a handle in subsequent operations.

Fig. 13-2. Aluminum pattern of top bow rest pad. A hole is drilled and tapped in the flat base for a 10-32 screw used as handle.

A good original part probably could have been used as a pattern had it been available, but rarely will an exposed rubber part be in good condition after fifty-plus years of exposure. It will almost surely be cracked, broken, or severely deformed. The pattern may also be made of rigid materials like hard wood, plastic, polyester body filler, or any other material that can be readily molded or carved.

MAKING THE PLASTER MOLD

The mold is made of Plaster of Paris. Use a container large enough to accommodate the pattern with room to spare all around—in this case a 5-ounce paper cup. Mix Plaster of Paris and water to a creamy, workable consistency and pour it nearly to the top of the paper cup. Plaster of Paris is available at art, craft, paint, or hobby stores. It is inexpensive. Always mix with water by adding the powder to the water—not the other way around. Mix only enough for immediate use because it sets in minutes and cannot be used again. The type of plaster used by artists should be used rather than the wall-patching variety found in builders' stores. The latter is usually much too coarse and may contain a retarder to increase the setting time.

You have achieved the proper consistency of the plaster mix when the material can barely be poured from one container to another. Too thin a mix will result in a weak mold. See Fig. 13-3 show-

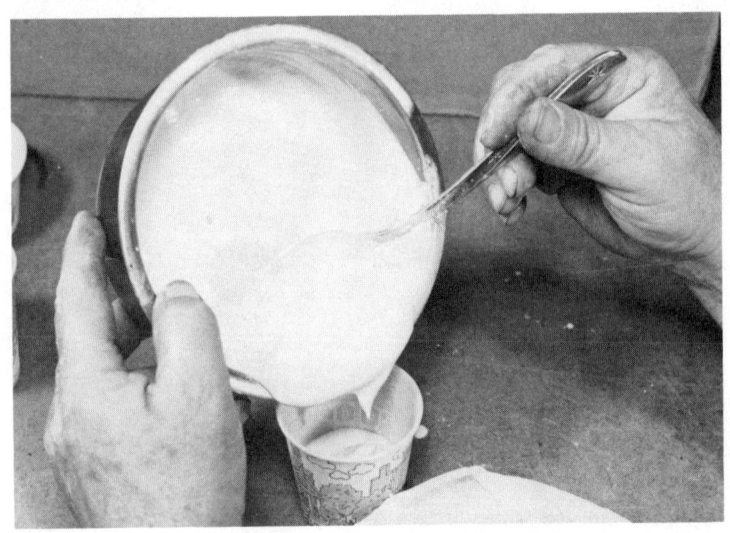

Fig. 13-3. Transferring the plaster mix to the paper cup.

Fig. 13-4. Placing the pattern in the wet plaster.

ing the transfer of the plaster to the paper cup.

Rub a thin layer of Vaseline on the pattern and press it gently into the wet plaster. Allow the pattern to sink of its own weight until the plaster oozes slightly above the level of the flat base of the pattern (Fig. 13-4). The mold should be allowed to set until it hardens slightly—about 15 minutes. Use the flat blade of a knife or spatula to carve the plaster down to the level of the pattern base. In another thirty minutes or more the plaster will completely harden and the pattern may be withdrawn from the mold.

WORKING WITH PLASTER

If you have never worked with Plaster of Paris before, you will quickly observe that it has some peculiarities that are difficult to describe in print, but the perceptive craftsman will quickly master the techniques of plaster molding. Plaster of Paris is made from the common mineral gypsum, a white, rocklike natural product composed of hydrated calcium sulphate. When gypsum is ground to a fine powder and heated to drive off some of the combined water, it becomes the useful product so widely used in plasters. When the powder is mixed with water, a chemical reaction occurs between the water and the calcium sulphate, and the material reverts to its original rocklike form of gypsum. The reaction takes place rapidly, so for purposes like plastering walls, retarders are added to slow down the setting rate.

Retarders are not essential for mold making. If only a slight retarding action is desired, ice water may be used for making the plaster mix. The final setting of the material is indicated when the mold becomes perceptibly warm to the touch. This is a normal consequence of the chemical reaction.

Discard surplus unused plaster by dumping it into a suitable container for disposal. Do not pour it down a sink because it will set up under water and may clog the plumbing. Since nothing readily dissolves gypsum, the problem could be severe.

When you mix Plaster of Paris with water some air inevitably gets into the mix. Large bubbles come to the surface, but small ones are not so easy to get rid of. The professional plaster molder will equip himself with a vacuum bell jar to deaerate his mixtures, but the amateur restorer usually will not have such equipment. Jogging the mix by tapping the container against the the table or work bench will dislodge many stubborn air bubbles, but this is only partially effective, and therefore you may often find some pits in the mold caused by air bubbles.

PREPARING THE MOLD FOR USE

After the plaster has taken the initial set and the pattern has been removed from the mold, examine the mold for small pits caused by air bubbles or other causes. Small pits can be filled with additional small quantities of wet plaster applied with the end of a curved spatula or similar tool. After the repair has set, it may be smoothed out with fine sandpaper held on the end of a wooden tongue depressor or similar tool. Pits may also be filled with body glazing putty, but only after the mold is thoroughly dried.

Dry the mold thoroughly. Air drying at room temperature may take several days, but if you want to speed up the process, place the mold in an oven at the lowest possible temperature setting (not over 200 °F.) for a few hours or place it in a gas oven with only the pilot light on. It should dry overnight. Too rapid drying at an elevated temperature may cause the mold to spall because of rapid evolution of steam from the wet mold. If you use the mold to make only one part, no further treatment is required. If you use the mold to make more than one part, or keep it for possible future use, give the cavity of the mold and the area around the cavity two or three thin coats of shellac. Allow about one-half hour drying time between coats.

The shellac seals and strengthens the somewhat porous plaster mold and provides a smooth molding surface. There may be other sealing materials that can be used for this purpose, but I have never found anything better than shellac. Shellac that is several years old will not dry properly, so be sure the stock is reasonably fresh. If it does not dry firm and hard, discard it.

POURING THE MOLD

Coat the mold's interior with Devcon release agent. If that is not available, apply a thin coating of silicone grease, Teflon spray, or automobile wax to prevent the Flexane from sticking to the mold. Mix the Flexane according to the manufacturer's instructions, or follow the directions given in Chapter 12: Rubber and a Space-age Substitute. If you mix less than the full one-pound unit quantity, it is necessary to measure the ingredients carefully. The pad de-

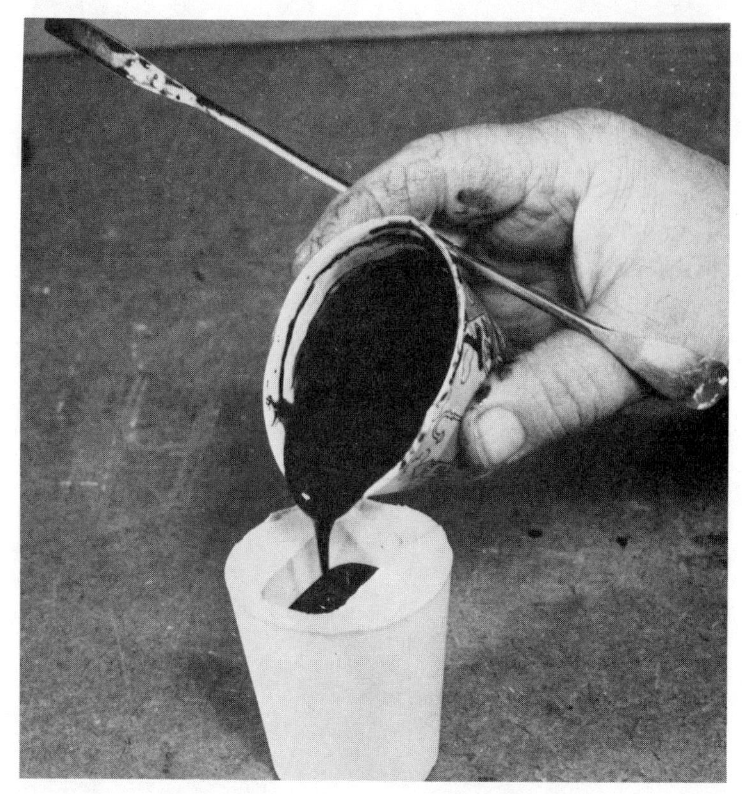

Fig. 13-5. Pouring Flexane polyurethane rubber into the mold.

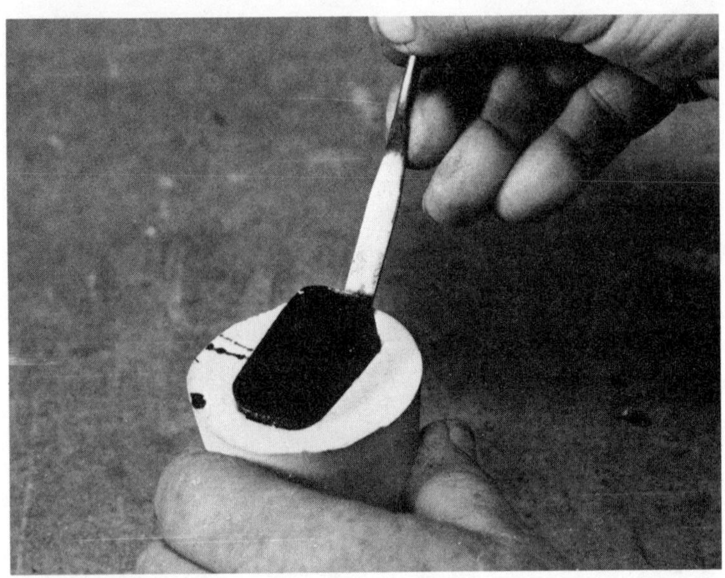

Fig. 13-6. After allowing the Flexane polyurethane rubber to cure for 8—10 hours, the part may be pried out of the mold.

scribed here weighs only about three-fourths of an ounce. It is impractical to weigh out such small quantities for mixing, so to avoid wasting material, pour other molds at the same time.

Slowly pour the Flexane into the mold in a small stream as shown in Fig. 13-5 so that air is forced out of the mold. Fill the mold level with the top. Allow it to set for at least eight hours. Then

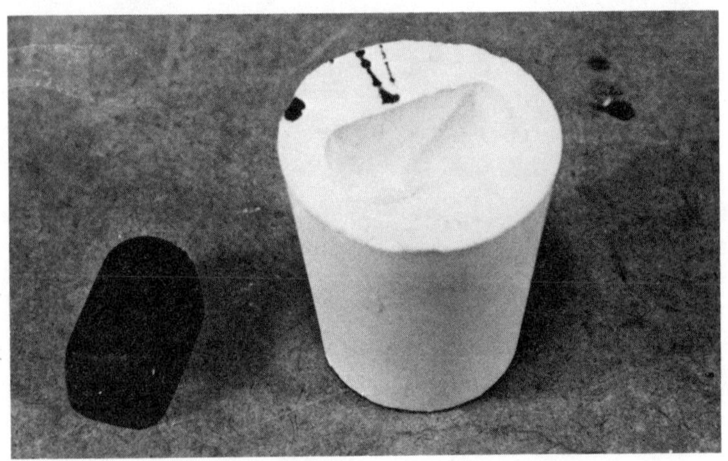

Fig. 13-7. The mold, right, and the top bow rest pad made in it.

you can remove the part from the mold by prying it out with a spatula blade as shown in Fig. 13-6.

You can make the pad with Flexane 80 liquid, but the pad will be rather hard. It is better to aim for a durometer of about 60 in this kind of part in which case add, Flex-Add to the Flexane 80 liquid in the proportion of six ounces of Flex-Add to one pound of Flexane 80. The reproduction pad and the mold used for making it are shown in Fig. 13-7.

Chapter 14

Making a Four-Hole Grommet

In this chapter we undertake the reproduction of a more complicated rubber part, a four-hole *grommet* made in a two-piece plaster mold. By gradually progressing to more difficult examples, you will find it easy to adapt the methods to your requirements and begin to develop your own designs and ideas. The possibilities are limitless.

Nearly every antique car has at least one feedthrough grommet on the firewall for control cables, choke control, speedometer cable, oil gauge line, and manifold heat control. Unless they are a simple circular pattern, replacements are not likely to be available. The grommet we will make is oval with four openings in it and apparently is not available commercially.

The original grommet on the car was intact, but was badly warped, deformed, and rock-hard—not suitable for use as a pattern. Most rubber parts from antique cars require some reworking to be used directly as patterns.

Dimensions taken from the original part were confirmed in part by comparing them with the dimensions of the hole in the firewall. The sketch in Fig. 14-1 was made and a pattern made of aluminum. The original grommet and the pattern are shown in Fig. 14-2. No holes were made in the pattern because it was planned to bore holes in the grommet after it was made. You can also make the pattern of wood, wax, polyester resin, or similar materials which can be readily molded or carved.

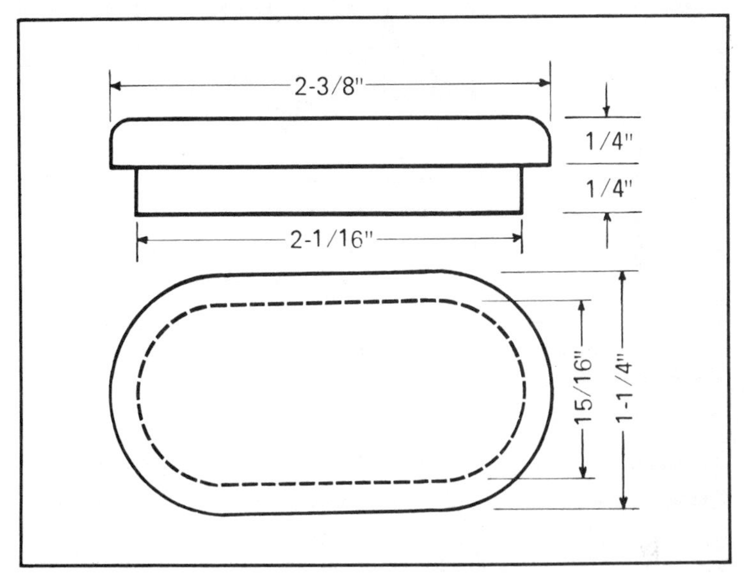

Fig. 14-1. Sketch of pattern for a firewall grommet.

MAKE A PLASTER MOLD

The mold is made of Plaster of Paris. Use a cardboard box that is about 2 inches deep and large enough to accommodate the pattern with room to spare. Mix Plaster of Paris with water to a workable consistency and pour it into the cardboard box so it is half full.

Press the pattern into the plaster before it sets as shown in Fig.

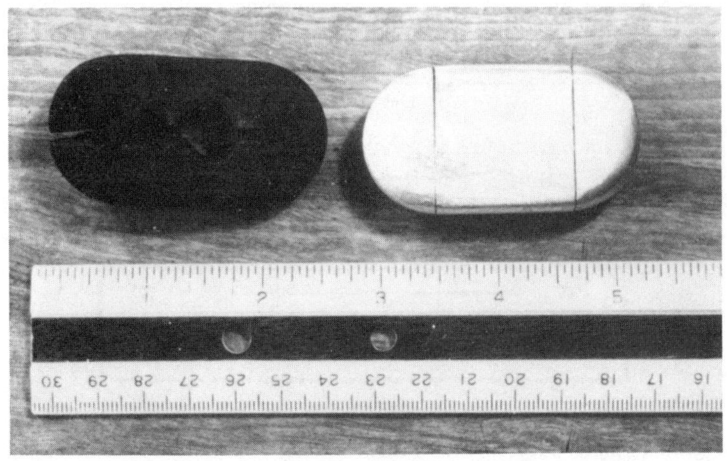

Fig. 14-2. Original grommet, left, and an aluminum pattern at right.

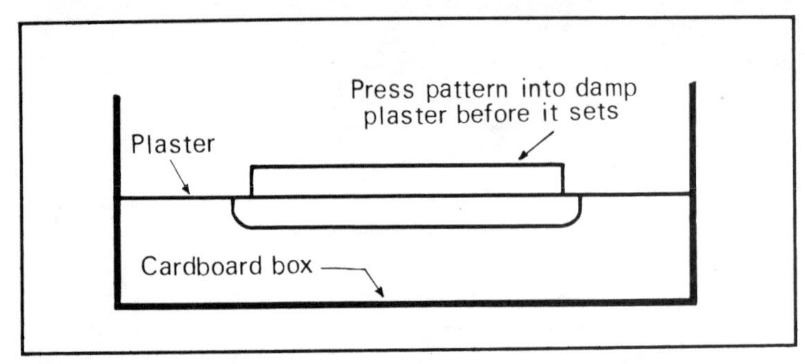

Fig. 14-3. Fill cardboard box or form about half full and press pattern into wet plaster as shown.

14-3. Rub a thin layer of vaseline on the pattern first so it will not stick to the plaster. After the plaster has set—usually in a half-hour—coat its surface and the exposed part of the pattern with a thin layer of vaseline. Add enough plaster to fill the box. You may gouge some shallow holes in the bottom half of the mold as shown in Figs. 14-4 and 14-5 to key the two halves of the mold together.

When the mold is completely set, pry the two halves of the mold apart and remove the pattern. You can fill in any small pits or imperfections in the mold surface with more plaster and smooth it out. For some hours after the initial setting reaction, the plaster will be fairly hard but it can still be carved and cut with sharp tools. After prolonged setting it becomes almost rock hard and it is very difficult to work. Before this occurs, carve a hole through the top of the mold and into the mold cavity. This hole should be at least one-half inch in diameter and it may be enlarged to a funnel shape at the top as shown in Fig. 14-4.

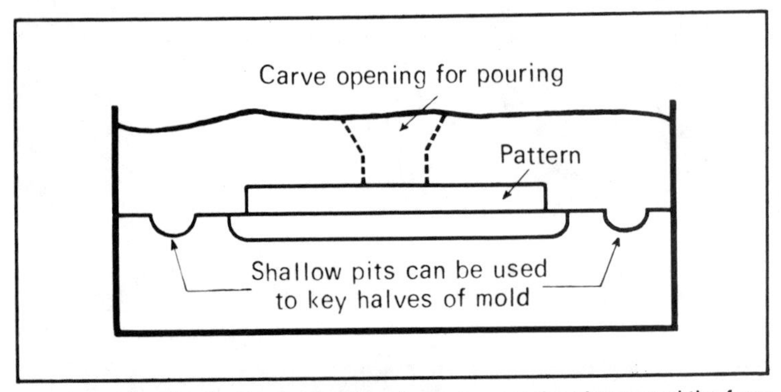

Fig. 14-4. After the first layer of plaster has set, coat the plaster and the form with petroleum jelly, and add more plaster.

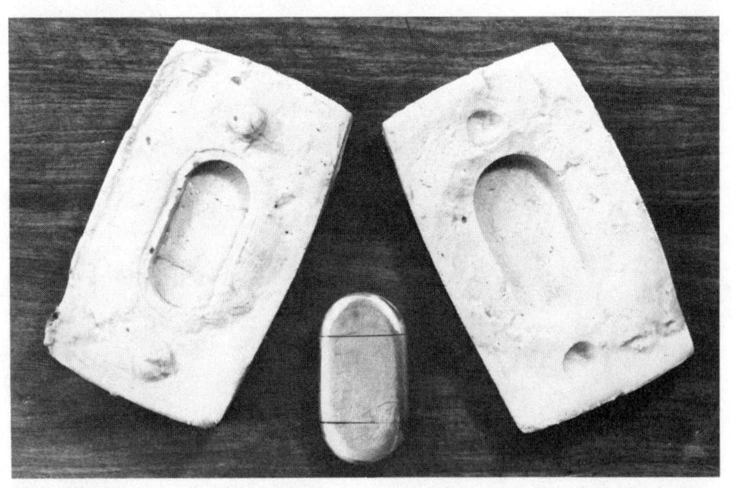

Fig. 14-5. The two halves of the mold have been separated and the pattern removed. A funnel-shaped opening will be carved into the upper half of the mold before it sets. Note the pits and projections at the ends of the mold that key the two halves together and insure perfect alignment when the mold is assembled.

If you have read *Part 1—Casting Metal Parts*, you can see the similarity between a plaster mold and a typical sand mold for metal casting. The cope, the drag, and the sprue all have counterparts in the plaster mold.

You may ask why the plaster mold cannot be used for pouring a metal part.

In fact, we could use the plaster mold just made for pouring metal if we modified it. In the form described, the mold would not be heat-resistant enough for casting brass, bronze, or even aluminum, but with a little more drying we could cast some low melting alloys in it. Plaster molds are used for casting, even for brass and bronze, but the plaster must be mixed with sand or talc to make it heat-resistant. The molds must be be baked at high temperature to remove as much water as possible. It takes more time to prepare plaster molds (not emphasized in Part 1).

The mold just described is a simple one. More complicated patterns may require that the mold be in more than two pieces. On the other hand, simpler shapes may be cast in a one-piece, open-top mold. It takes ingenuity to design molds, but once the basic principles are mastered you should have no problem developing techniques to fit desired applications.

After you complete the mold, allow it to air dry for at least two or three days before you use it. If you need to speed up the drying,

Fig. 14-6. The mold and the rubber part cast in it.

the mold may be put in an oven at the lowest possible heat—preferably not over 200 °F.—for several hours. If you want to use the mold to make only one part, no further preparation of the mold is necessary. If you plan to use the mold in the future for making several pieces, give the cavity of the mold and the faces where the two halves of the mold touch two or three thin coats of shellac.

PREPARING THE MOLD AND THE MIXTURE

Prepare the mold for pouring by coating the interior surfaces with a release agent to prevent the rubber from sticking to the mold cavity. Mix the Flexane by weighing the proper proportions of resin and hardener according to the manufacturer's instructions. Fairly careful measurement is necessary, so use a small chemical balance, postal scale, or photographic scale. Measuring the materials by volume is much less desirable. Mix only enough for immediate use as the material has a pot life of only 20-30 minutes after the resin and hardener are mixed. If you use the entire contents of the kit at one time, no measuring is necessary. Small paper cups are convenient for measuring and mixing small portions. The resin and hardener must be thoroughly blended together, but too vigorous stirring will cause too many air bubbles to be drawn into the mix.

Slowly pour the Flexane into the mold in a small stream so that all the air can escape from the mold. Fill the mold so that the mixture comes slightly up into the filler opening. The projection can be trimmed off later.

Allow the mold to rest undisturbed for at least eight hours for the rubber to cure. Then separate the mold and remove the part as shown in Fig. 14-6. Complete cure will take a couple more days.

All that remains is to cut off the plug from the pouring opening and bore the appropriate holes in the grommet. The holes were bored with sharpened pieces of brass tubing rotating in a drill press. The result is a beautiful reproduction of the original rubber part. This part was made with Devcon Flexane 80 Liquid which produces a medium hard rubber.

Chapter 15

Making Door Bumpers and Check Straps

Almost every car restorer faces the vexing problem of replacing the little rubber bumpers found in automobile door jambs. The bumpers are usually good for many years, but on most of the older cars they have flattened out or become rock hard with age. Either way, they can no longer perform the small but vital function they were designed for—cushioning the door so that it does not contact the jamb and cause rattles.

Many door bumpers are block or wedge-shaped, but there are as many different styles, shapes, and sizes as there are car models. Sometimes it is difficult to find the exact replacement. A couple of typical bumpers were selected from the junk box for practice reproduction using polyurethane rubber. Both were slightly the worse for wear, but they appeared to be in good enough shape to use as patterns. A section was missing from one bumper, probably a result of repeated contact with the door. The missing section was filled in with silicone rubber which was smoothed to the original contour of the part.

Plaster of Paris molds were made by the same procedures given in the preceding chapter. The accompanying pictures show the pattern, molds, and reproduction parts.

MAKING A CHECK STRAP PATTERN

Most antique cars use door check straps made of fabric-reinforced rubber several inches long. One end is attached to the

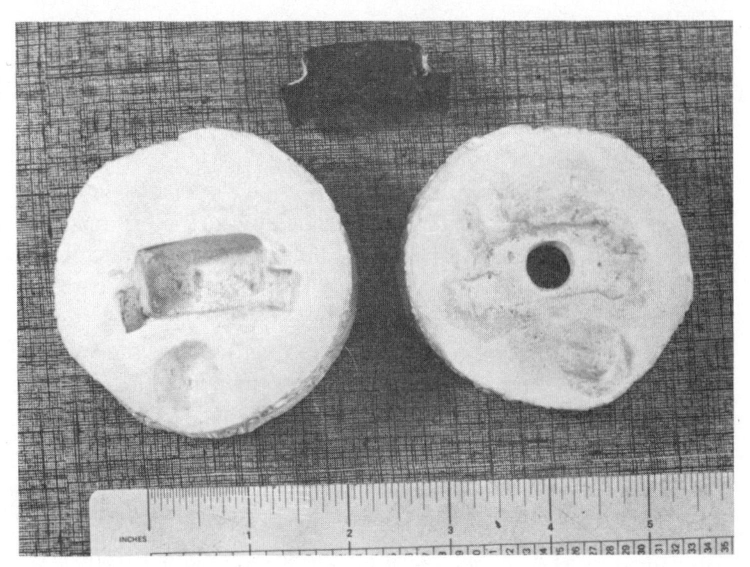

Fig. 15-1. Two-piece plaster mold for a small door bumper. The original part was used as a pattern.

edge of the door and the other end to the door hinge pillar. The function of the check strap is simple—it limits the distance the door can be swung open to prevent it from striking some other part of the car. Modern cars usually use a more complicated spring-loaded metal device.

A particular type of door check was needed for replacement

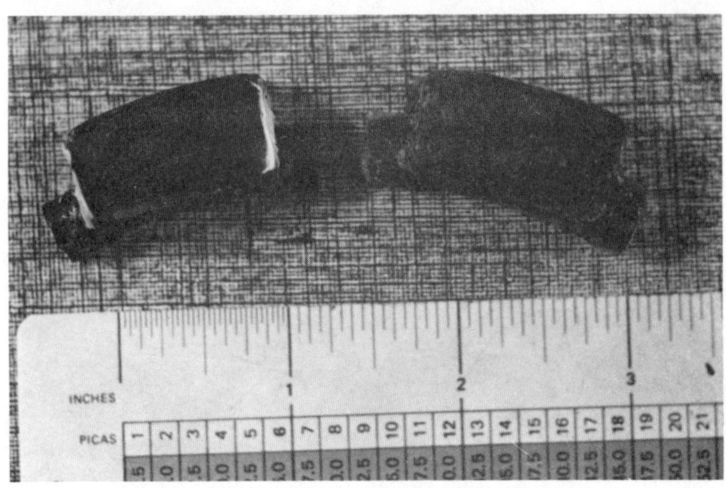

Fig. 15-2. Original part, left, with reproduction part at right.

purposes. It consisted of a typical reinforced strap about 6 1/2 inches long with a piece of metal strip molded into an enlarged section at one end. In use, the free end is attached to the edge of the door with four sheet metal screws, and the T-shaped metal strip on the other end catches inside a slot in the door hinge pillar when the door is in the fully opened position. The best original door check was selected as a pattern, but even though efforts were made to build up worn sections with silicone rubber, the check was too far gone. Therefore, a wooden pattern with the exact dimensions of the original part was made as shown in Fig. 15-3. The pattern was sprayed with about five coats of clear lacquer to make the surface smooth and moisture proof. After coating the pattern with a thin layer of vaseline, a two-part plaster mold was made using the same technique as outlined before.

REPRODUCING DECORATIVE GROOVES IN RUBBER

The original strap has a system of fine longitudinal grooves on each side, apparently purely a decorative feature. A piece of hacksaw blade was ground to fit the contour of the depression in the mold. Teeth on the extreme ends were ground off. Suitable

Fig. 15-3. The original door check strap, top. The white areas are the result of a futile attempt to fill cracks and abraded areas with silicon rubber. When this plan failed, the wooden pattern, below, was made.

Fig. 15-4. The plaster mold made with the wooden pattern.

Fig. 15-5. A piece of hacksaw blade was drawn through the mold to form a system of parallel grooves to simulate the design on the original part.

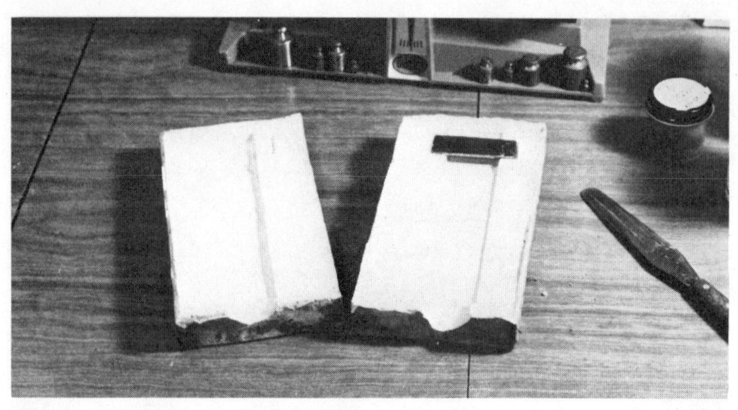

Fig. 15-6. The mold ready to assemble with the metal insert in place.

shallow grooves were scratched in the mold surface by drawing the blade through the mold. The process was repeated for the other half of the mold as shown in Fig. 15-5.

A small piece of steel was cut to the correct size and laid in the transverse slot in the mold as shown in Fig. 15-6. The halves were assembled and clamped together with sturdy rubber bands after the mold cavity was brushed with a release agent. The mold was poured with Devcon Flexane 80 Liquid with the mold standing on end, and the t-shaped end down as shown in Fig. 15-7. No cloth reinforcement was used in the reproduction part, and none is needed

Fig. 15-7. Pouring the polyurethane rubber into the mold. The latter is standing on end and held together with rubber bands. Flexane 80 Liquid polyurethane rubber is recommended for this part.

Fig. 15-8. The reproduction part as it appears when the mold is opened.

apparently because of the enormous tensile strength of polyurethane.

The rubber reproduction exercise in this chapter illustrates: (1) producing a composite part of rubber and steel; (2) use of a wooden pattern; and (3) modifying a mold to produce a decorative pattern on a part. These are a few examples of the versatility of the rubber casting process.

Chapter 16

Making a Fender Lamp Pad

This reproduction exercise is a radical departure from techniques described earlier. Instead of pouring the rubber into plaster molds, the parts described will be made in open wooden molds. The technique is well suited to making thin, flat parts like mats, gaskets and pads. How to produce a fender lamp mounting pad will be discussed. It is a flat piece approximately 1/16 inch thick with a one-eighth inch rounded bead around the perimeter. Its function is to prevent the lamp from marring the fender paint while the rounded bead prevents moisture from getting under the lamp.

The first step is to make a tracing of the outline of the lamp on a piece of cardboard as shown in Fig. 16-1. The outline could not be transferred directly to the board used for the mold because of the lamp's curvature. Next the cardboard pattern is cut out and its outline traced on the board as shown in Fig. 16-2. A groove is cut along the line using a 1/8 inch viening router bit in a drill press. The board has to be guided slowly and carefully to stay on the line as shown in Fig. 16-3, but the trick is easily mastered with a little patience.

RECESSING THE OPEN WOODEN MOLD

The next step is to cut out all of the wood inside the groove to a depth of 1/16 inch. You can use a router bit or end mill in a drill press for this operation, and Fig. 16-4 shows the use of a 3/4-inch end mill in a drill press with a board guided by hand. When

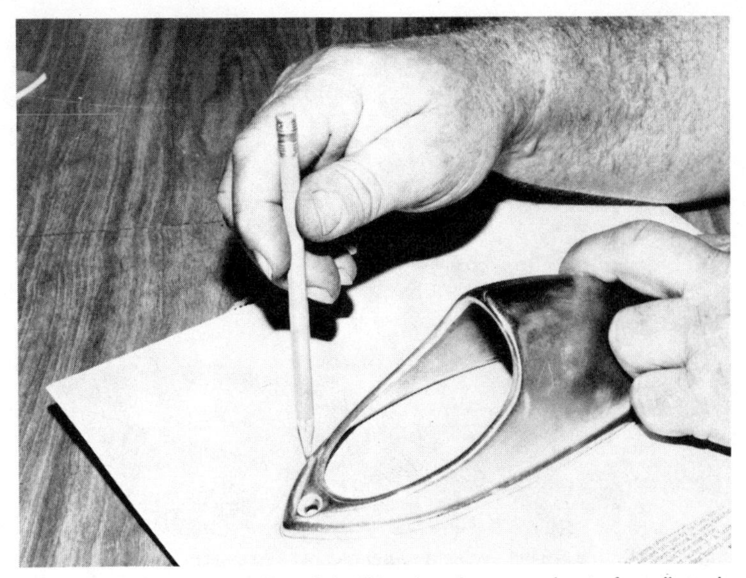

Fig. 16-1. Tracing the outline of the lamp housing on a sheet of cardboard.

finished, the typical cross-section of the mold in the vicinity of the groove should look like the sketch in Fig. 16-5. Tolerances are not tight and a few thousands of an inch one way or the other should make little difference.

FINISHING

Any rough spots in the mold should be sanded smooth.

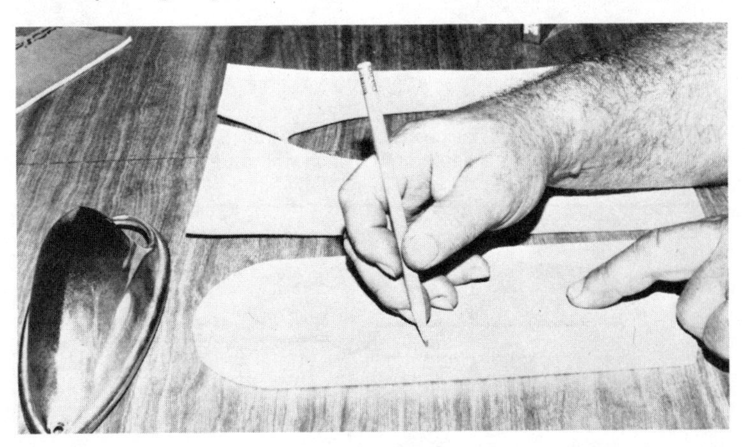

Fig. 16-2. The cardboard pattern is cut out and its outline is traced on a board. A piece of birch wood was used, but any other smooth hardwood is suitable.

Fig. 16-3. With a 1/8-inch veining router bit chucked in a drill press, the board is guided by hand to cut a 1/8-inch deep groove along the traced line.

Fig. 16-4. After the groove has been cut along the traced line, all material within the groove is cut out to a depth of 1/16-inch with an end mill or router.

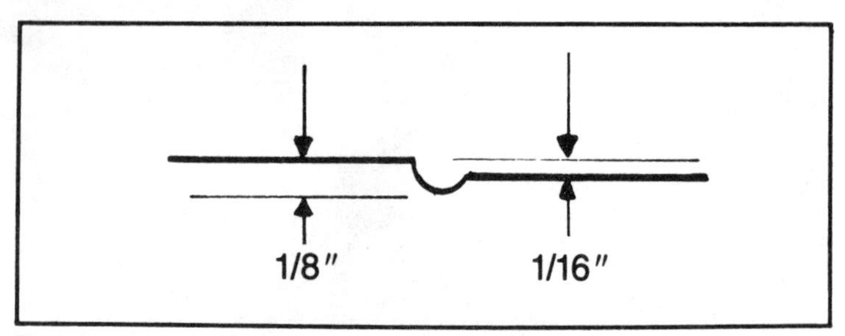

Fig. 16-5. A cross section of the mold near the groove will look like this.

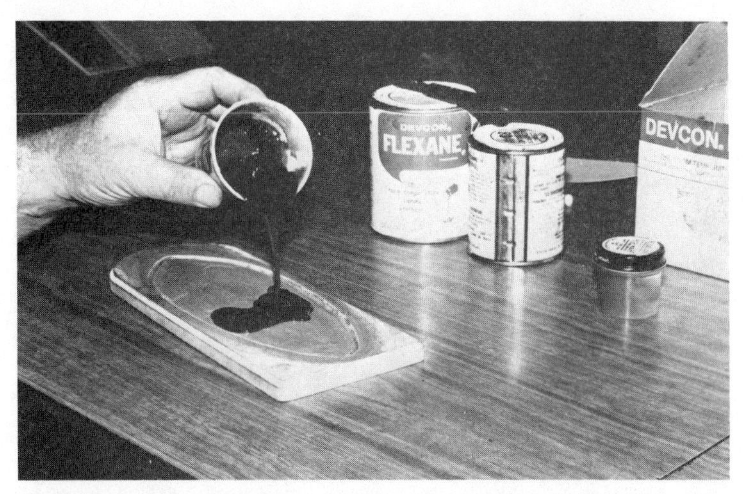

Fig. 16-6. Pouring the Flexane polyurethane rubber into the mold. A brush or spatula may be used to smooth the material to a uniform depth.

Smoothness in the groove is particularly important because only the bead shows outside the lamp base. Apply a couple of coats of clear lacquer or shellac to the mold, smoothing the mold with fine sandpaper between coats. After release agent is applied to the mold, it is ready to receive polyurethane rubber. Flexane 80 Liquid should be used. After pouring on the Flexane as shown in Fig. 16-6, it may

Fig. 16-7. The part may be peeled from the mold after curing.

Fig. 16-8. The finished part with the mold.

be smoothed out uniformly in depth using a brush or spatula. Allow about eight hours for partial cure before peeling the part out of the mold. Additional parts can be turned out in the same mold quickly and easily.

The finished part is peeled from the mold in Fig. 16-7. Figure 16-8 shows the finished part.

Chapter 17

Metal Molds for More Precision

The molding technique in this chapter differs from those described earlier because the part requires a higher degree of dimensional accuracy. The part to be made is a Pitman arm bushing, Chevrolet part number 599643. Four of these parts are used on all Chevrolet passenger cars from 1939 through 1948, except for 1939 and 1940 Master "85" with conventional front axle. These parts should still be available from General Motors parts suppliers but delivery is sometimes very slow. It can hardly be recommended that anyone reproduce a commercially available part but the exercise illustrates some important techniques and principles.

It is difficult to describe the function of the Pitman arm bushing in detail, but it is made up of two parts held together by a pair of studs and nuts. The studs are cushioned with the rubber bushings so that the Pitman arm has a small amount of flexibility to absorb steering shocks that would otherwise be transmitted to the steering wheel.

PREPARING TO MAKE A METAL MOLD

Preliminary inspection of the part indicated that a metal mold would maintain the relatively high order of dimensional tolerance that is necessary for the part. The first step was to carefully measure an unused part and make a sketch as shown in Fig. 17-1. Taking precise measurements from a rubber part with micrometer calipers is not easy, so the dimensions shown are rounded off to

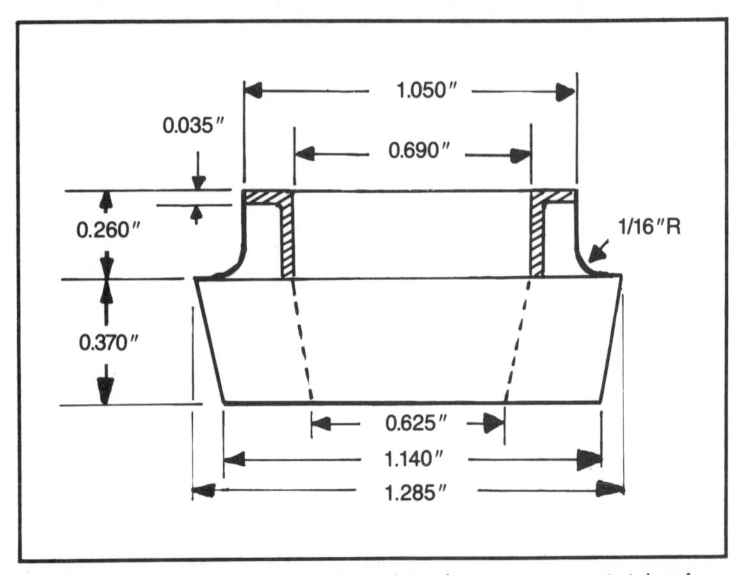

Fig. 17-1. Sketch of Pitman arm bushing based on measurements taken from an original part. The shaded part is a steel bushing.

Fig. 17-2. Making the mold out of aluminum stock. A lathe is needed to make this mold. It is made in three pieces for easy assembly and removal of the finished part.

Fig. 17-3. Machining the tapered plug that forms the core of the mold.

the nearest 0.005 inches which should be precise enough for this part. Note that the part includes a metal ferrule (shown as the shaded section in Fig. 17-1) bonded to the rubber.

The next step is to machine a mold out of three pieces of aluminum so that the mold can be disassembled to remove the part after it is cast. From this point on, the story can best be told in Figs. 17-2 through 17-10.

Fig. 17-4. The finished mold. The locating pins on the edge of the mold keep the two major parts aligned when it is assembled.

Fig. 17-5. The three-piece mold assembled.

Fig. 17-6. Checking the fit of the original part in the mold.

Fig. 17-7. This view shows the partially assembled mold with the metal ferrule in position. All of the mold's internal parts are coated with release agent, except the ferrule. The rubber must bond to it.

Fig. 17-8. Pouring the mold. A fairly hard rubber is required, so the part is made with Flexane 94 Liquid polyurethane rubber. The mold is filled to the level of the top of the central plug. The plug has been cut to the correct length to serve as a reference point.

Fig. 17-9. After the rubber is cured, the mold is disassembled and the part removed.

Fig. 17-10. The reproduction part compared to the original, right.

Chapter 18

Ingenuity Replaces Original Part

The rubber project in this chapter involves a highly specialized application for one make of car. Therefore, while relatively few readers are likely to be interested in this car part *per se*, the description of its reproduction illustrates what an amateur can accomplish with some ingenuity and thought. The parts bear the rather unwieldy names "headlight tie rod thrust washer, inner and outer," and are part numbers 252368 and 252369, respectively, for several models of Studebaker and Rockne cars built between 1931 and 1933.

The story behind the thrust washers is interesting. When Studebaker introduced its 1931 models, the fender cross bar that had been a feature of all its previous cars was eliminated. This omission proved to be a mistake because an alarming vibration of the front fenders sometimes occurred on rough roads. Studebaker fixed the problem by introducing a running change in its 1931 models. A horizontal tie rod was placed inside the radiator shell just in front of the radiator core. On each side, short rods ran from the terminus of the tie rod to the headlight brackets.

Any rigid connection of this short tie rod to the radiator shell resulted in localized stresses caused by vibration. Therefore, the end of this rod had a ball and socket joint cushioned with rubber thrust washers. The same stress conditions apply today, and some restorers have learned, to their dismay, that rigid bolting of the parts will cause stress cracking in the radiator shell.

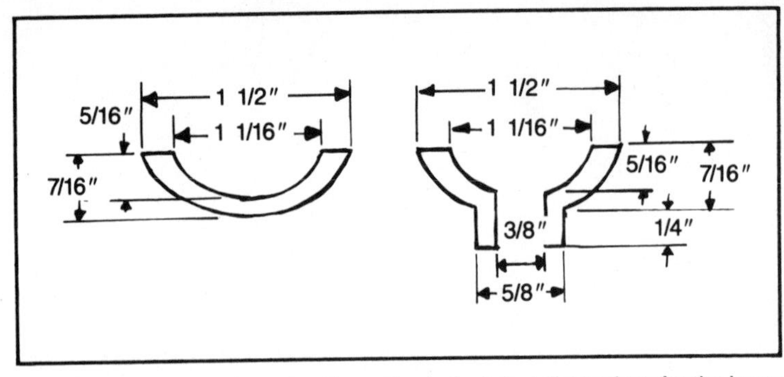

Fig. 18-1. These sketches give the estimated original dimensions for the inner and outer tie rod thrust washers.

ESTIMATE WITHOUT AN ORIGINAL

As with the Pitman arm bushing project, the first step was to make a dimensional drawing of the parts to be produced so that proper molds could be made. In this case, no intact original parts could be found, so the drawing in Fig. 18-1 represents the estimated dimension based on the measurements of the ball end of the tie rod and its mating parts. The dimensions may not be precisely the same as those of the original parts, but they should be close enough for practical purposes.

MOLDS

From the dimensional drawings, two molds were made of

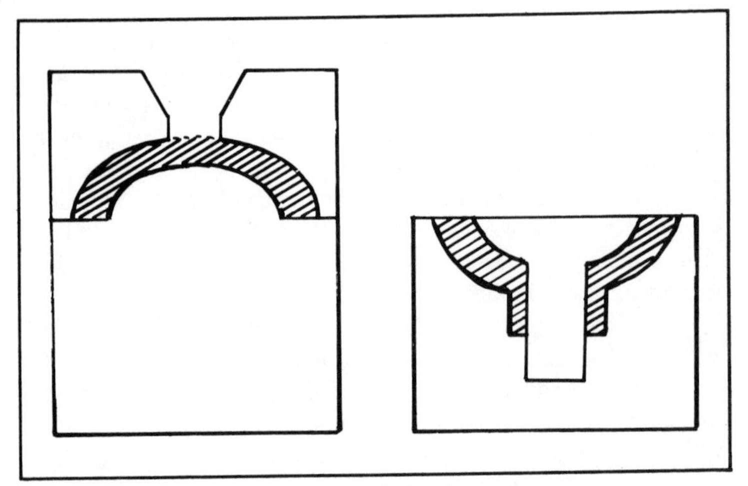

Fig. 18-2. The mold designs for reproducing the parts.

Fig. 18-3. The finished molds made from aluminum stock on a lathe. Wooden molds made with a woodturning lathe would be equally satisfactory.

aluminum stock as shown in Fig. 18-2. Similar molds could be made of wood turned on a lathe and shellacked before they're used.

Figures 18-3 through 18-7 show the molds' features, assembled and disassembled, and pouring the parts with Flexane 80 Liquid.

Fig. 18-4. The assembled molds.

Fig. 18-5. Pouring the mold for the inner tie rod thrust washer. The two parts of the mold are held together with a strip of masking tape.

Fig. 18-6. Pouring the mold for the outer tie rod thrust washer. When filling small molds like this one with minimum clearance between mold elements, it helps to tilt the mold slightly and pour from one side to displace the air.

Fig. 18-7. The completed rubber parts removed from their molds.

Chapter 19

Molding from Defective Patterns

The rubber project in this chapter involves the reproduction of a part commonly used on many makes of cars—the steering column collar. The part is often referred to as a steering column grommet. Its function is simple—to seal the opening between the steering column and the toe board.

No exact copy seems to be available commercially, but you can buy a variety of circular collars. A circular collar will work, but is not strictly authentic for a car model with an oval type.

The original circular collar on the car is in fair shape. It probably could be used as is, but it is hardened with age and a small section has been broken off along one edge, probably where the bottom of the clutch pedal has been striking it for years.

Missing sections of the parts that are to be used as patterns can sometimes be filled in with silicone rubber. In this case, the circular collar will be used as is for a pattern. Reworking the mold will correct the missing section—a practical demonstration of still another useful technique.

With this information as a background, the rest of the story is told in Figs. 19-1 through 19-12.

Fig. 19-1. The original part—hardened with age and with a section broken from the edge, but still good enough for use as a pattern.

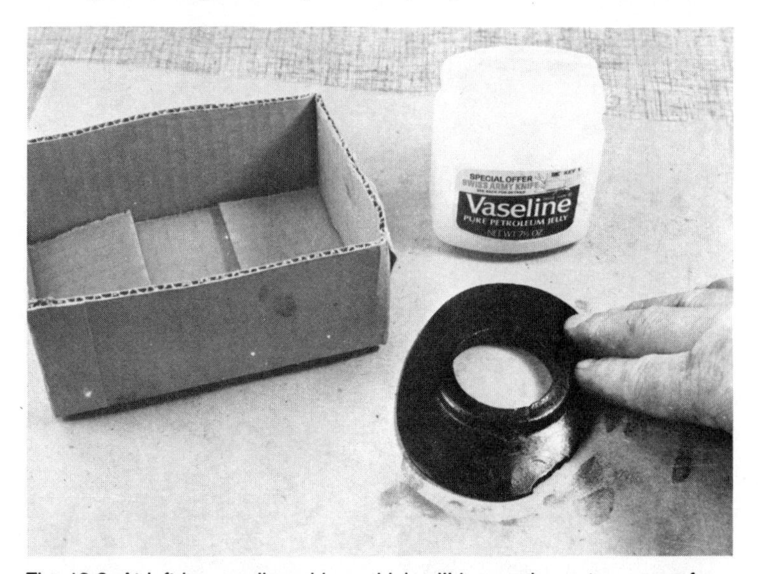

Fig. 19-2. At left is a cardboard box which will be used as a temporary form for the plaster mold. The pattern is coated with petroleum jelly to prevent the plaster from sticking to it.

Fig. 19-3. The box is filled approximately half full with plaster mix.

Fig. 19-4. Before the plaster begins to set, the pattern is placed on the surface and pushed down until it is level with the plaster's surface.

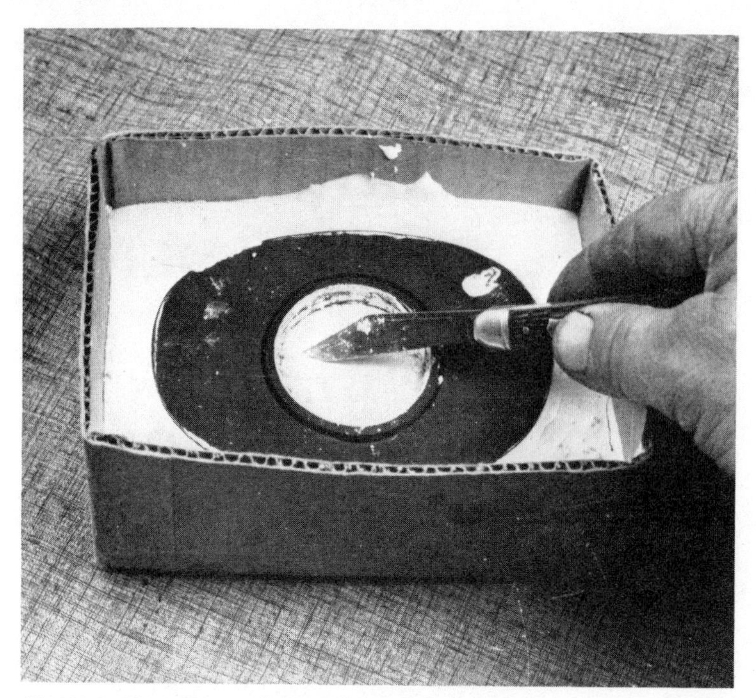

Fig. 19-5. After the mix has partially set, the plaster that has oozed into the central section is scraped away.

Fig. 19-6. The pattern and exposed surfaces of the plaster are coated with a thin layer of petroleum jelly.

Fig. 19-7. Another batch of plaster is mixed and the box filled near its top.

Fig. 19-8. After the plaster is thoroughly set (about 30 minutes), the cardboard form is peeled away from the mold.

Fig. 19-9. The halves of the mold are pried apart and a pouring opening is carved into the upper part. The defect in the mold caused by the missing section of the pattern can be corrected by using a sharp knife to carve the plaster to the correct contour of the mold. Other defects in the mold can be filled and corrected at this time.

Fig. 19-10. The completed mold should be dried thoroughly before use. It may be air dried for several days or placed in an oven (not over 200° F.) for a few hours. If the mold will be used in the future, it should be given two or more coats of shellac.

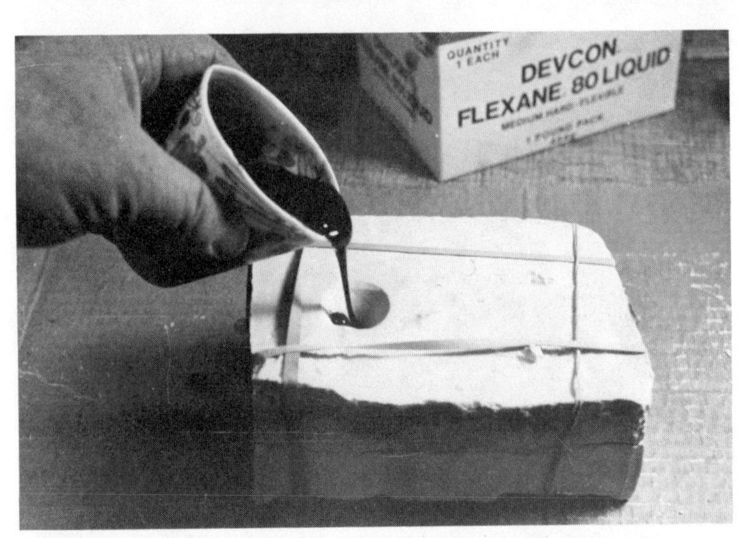

Fig. 19-11. Pouring the mold. A Flexane polyurethane rubber with a durometer hardness of from 60 to 80 may be used for a part of this kind. The inner parts of the mold must be coated with release agent before pouring.

Fig. 19-12. The reproduction part as removed from the mold.

Chapter 20

Molding with a Metal Insert

Hood corners will probably not be high on the car restorer's list of priority items. There is such a wide variety of them available from commercial sources that it should be easy to find one to fit nearly every antique motor car. The reproduction of a hood corner is a fitting project to illustrate the production of a part with a deep slot or recess formed by use of a mold insert. The detailed procedures are shown in Figs. 20-1 through 20-9.

Fig. 20-1. The first step is to make a pattern of the hood corner. Two triangular-shaped pieces and a spacer, (far left), have been cut from 1/8-inch plywood. The two pieces, right, are right isosceles triangles but the camera angle has distorted the shapes. The 1/8-inch-thick pattern insert has been cut from aluminum stock. It simulates the shape of the corner of the hood.

Fig. 20-2. The wooden parts of the pattern are glued together.

Fig. 20-3. The rough pattern and the metal insert.

Fig. 20-4. The corners and edges of the wood pattern have been rounded and contoured by sanding. This view shows how the metal insert fits snugly inside the wooden pattern.

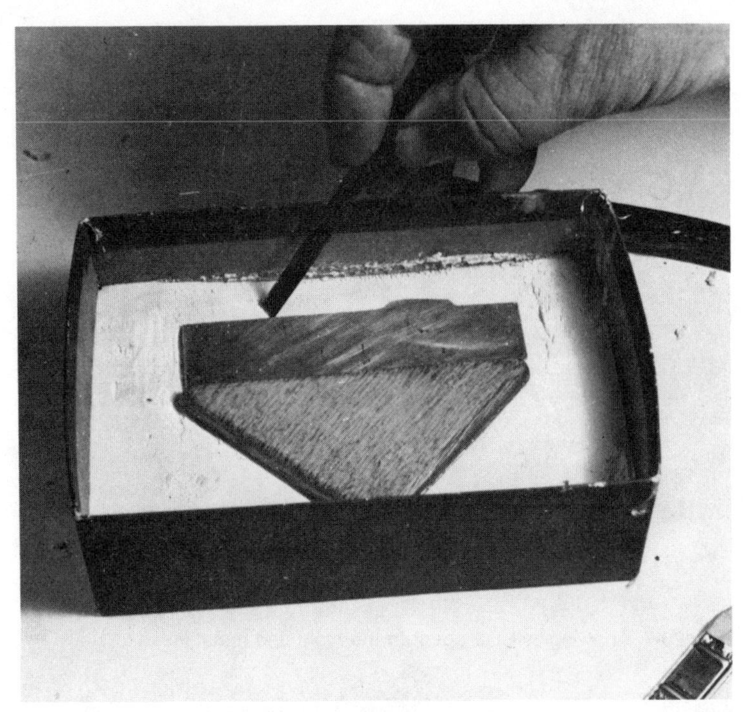

Fig. 20-5. A plaster mold of the assembled pattern and insert is made. The short projection on the edge of the metal insert assures that later it will be correctly assembled in the mold.

Fig. 20-6. The finished plaster mold with wooden pattern removed and the metal insert in position.

Fig. 20-7. Applying release agent to the mold and metal insert.

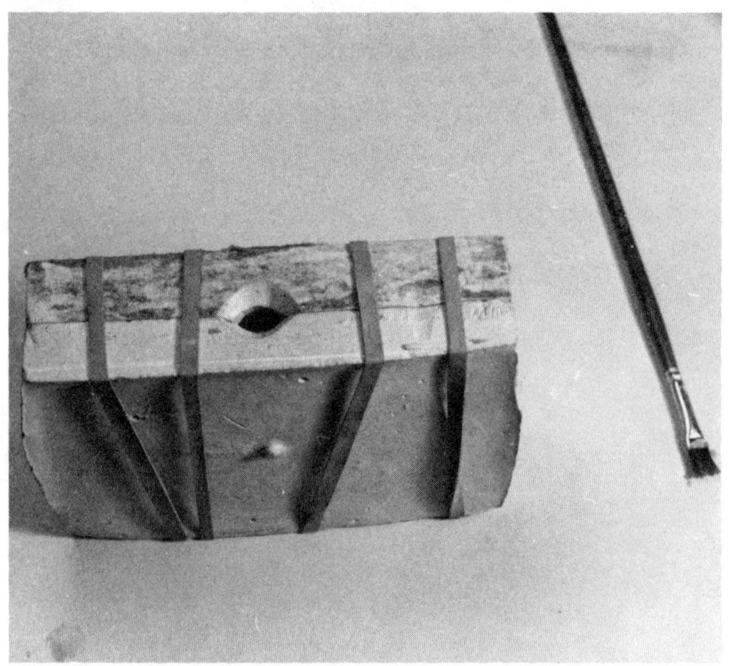

Fig. 20-8. The mold positioned to pour with the apex of the triangle at the top.

Fig. 20-9. The disassembled mold with the finished part removed.

Chapter 21

Making a Weatherstripping Mold

Nearly every automobile has weatherstrips to seal doors, windows, and windshields. A considerable number of commercial products are available, but the exact shape needed for restoring an antique automobile often can't be found. In this case it is feasible to cast a weatherstrip using a wooden mold. Simple geometrical shapes are easy to handle.

The weatherstrip project outlined is for a simple design—an L-shaped piece as shown in cross section in Fig. 21-1. A 44-inch-long piece of this shape was needed to seal the bottom of the windshield on an antique car. The mold was made by cutting a groove of the proper dimensions in a piece of white pine 1 × 2 lumber. A groove of the required shape could be cut with a router or by multiple passes on a table saw. We used a milling cutter because metal working tools were at hand. An end mill chucked in a drill press was used to cut a groove 3/16-inch wide and 3/8-inch deep, along the full length of a 48-inch long 1 × 2 board. Figure 21-2 shows the beginning of the cut with a board clamped to the drill press table as a guide. Figure 21-3 shows the board part way through the cut as the board is fed by hand past the cutter.

The first milling cutter was replaced with a 1/2-inch end mill to cut the rest of the groove to a depth of 1/4 inch, overlapping the first cut just enough to produce the desired total width of 9/16-inch as shown in Fig. 21-4.

The finished mold is shown in Fig. 21-5, the interior was given

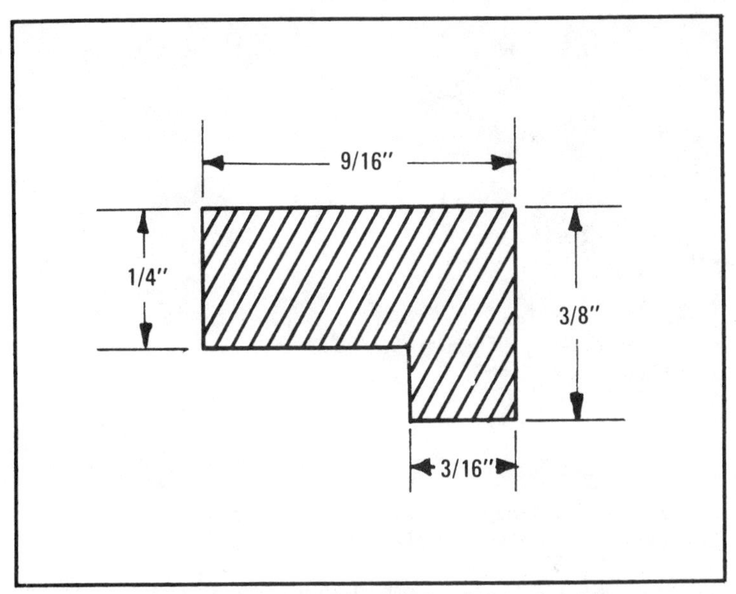

Fig. 21-1. Cross section of an L-shaped design for weatherstrip.

Fig. 21-2. Starting the first cut with a 3/16-inch end mill set to cut to a depth of 3/8-inch.

Fig. 21-3. Part way through the cut as the board is fed by hand. A high spindle speed should be used for smooth cutting action.

a couple of coats of shellac with sanding between coats to seal and smooth the wood.

Now it is simple to coat the interior of the mold with release agent, lay the mold horizontally on a level surface, and fill the

Fig. 21-4. Making the second overlapping cut with a 1/2-inch end mill.

Fig. 21-5. View of the end of the mold showing the shape of the finished groove.

groove with polyurethane rubber. Devcon 60 was selected for this part because a soft rubber was required. Figure 21-6 shows the beginning of the pour. The open ends of the mold have been dammed with strips of masking tape to keep the rubber from running out.

It behooves you to work fast with such a lengthy mold. The pot life of Flexane lasts 20-30 minutes after mixing, but the material usually gets more viscous after about ten minutes, so it is good to complete the pour in this time. If necessary, use a spatula or

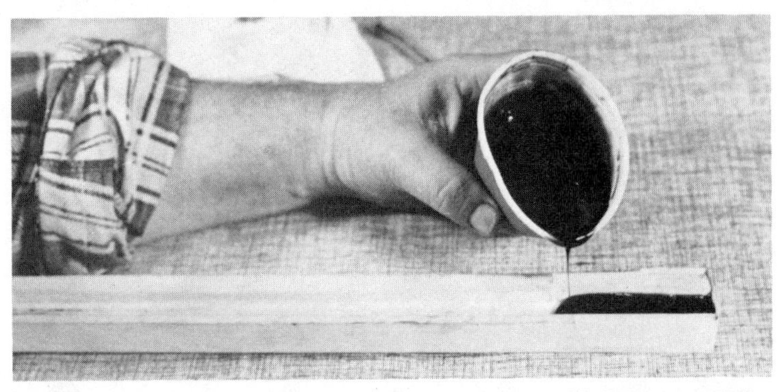

Fig. 21-6. Pouring the Flexane polyurethane rubber to fill the groove to the top of the mold. The open ends of the mold are dammed with tape.

Fig. 21-7. After the Flexane polyurethane rubber has cured, the weatherstrip is peeled from the mold.

knife blade to spread the Flexane along the groove, but usually the material will flow enough to level itself out nicely.

Figure 21-7 shows the finished product as it is pulled out of the mold after the Flexane has cured about eight hours.

Chapter 22

How to Estimate Amounts Needed

Flexane is expensive material, and once the curing agent is added to it, the mixture must be used at once. Any excess is wasted. It behooves one to accurately estimate the amount that is going to be needed to fill a mold to avoid undue waste. If you must guess at the amount, it is better to underestimate than to overestimate because you can mix a little more to finish filling the mold.

If the mold cavity is a regular geometrical shape, its volume can be accurately calculated. As an example, let us calculate the amount of Flexane needed to make a part with the dimensions given in the sketch in Fig. 22-1.

Convert fractions to decimals and calculate the volume of the sections marked A and B.

Vol. of A = 0.562 × 0.25 × 48 = 6.74 cubic inches.

Vol. of B = 0.125 × 0.187 × 48 = 1.12 cubic inches.

Total volume = 7.86 cubic inches

Flexane occupies a volume of 26 cubic inches per pound, so 7.86/26 = 0.30 pounds of Flexane will fill the mold.

Allow 10 or 15 percent extra for mixing losses and measurement error and mix 0.35 pounds.

If the mold is not a regular geometrical shape, it may be difficult or impossible to calculate its volume. In this case, several alternate procedures may be considered.

1. If an original rubber part is used as a pattern, weigh it and use the same weight of Flexane mix to fill the mold. Most rubber

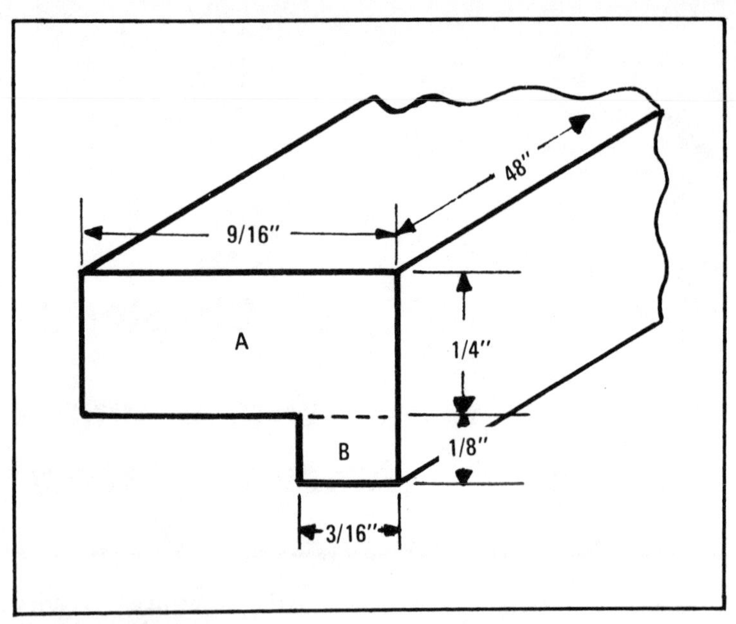

Fig. 22-1. Using dimensions to calculate the amount of Flexane polyurethane rubber, required to fill a mold.

parts are about 10 percent denser than Flexane, so if the same weight is used, an overrun of about 10 percent will compensate for mixing and transfer losses.

2. The mold can be filled with water and its volume measured after transferring the water to a suitable measuring vessel. For each ounce of water measured in this manner, allow 1.1 ounces of Flexane plus an added allowance for mixing and transfer loss. Water should not be poured into a plaster mold because the plaster will absorb it. If water is used in any other type of mold, dry the mold before use.

3. If water cannot be used, sand, salt, or sugar can be substituted. Transfer the granular material to a measuring vessel as with the water method above. The volume of the material in fluid ounces is multiplied by 1.1 to determine the required weight of Flexane. Again, a 10 percent extra allowance should be made for mixing and transfer losses.

Chapter 23

Be Careful, Clean, Dry, and Accurate!

Like any other chemical substances, the ingredients in polyurethanes may cause allergic reactions in some individuals. Proper care should be taken in handling the materials. If they come in contact with the skin, they should be removed with soap and water. Keep the materials out of the reach of children and pets, and provide adequate ventilation when handling them.

It is essential to clean all tools used for mixing and handling immediately after use. Several solvents can be used, but acetone, methylethyl ketone (MEK), and lacquer thinner are particularly useful. Exercise care as all these solvents are extremely flammable!

For weighing and mixing it is convenient to use paper dishes or cups or plastic cups.

Keep all containers tightly closed until immediately before use. Close containers tightly after use. Unopened containers have a shelf life of at least one year.

Flexane will not stick to oily, greasy, or dirty surfaces. For maximum adhesion to surfaces, clean thoroughly and roughen slightly, if possible. If you do not want Flexane to stick to a particular surface, such as the interior of a mold, apply release agent. Silicone grease or Teflon sprays are effective release agents.

For a complete cure, Flexane must be at a temperature of at least 60 °F. when it is applied or poured into molds. Flexane putties are mixed as liquids, but within five minutes the mixture sets to a non-sagging putty. Flexane liquids remain liquid for at least

15 minutes after mixing with the curing agent. If it is to be poured into a mold, the mixture must be used during this period of time.

Accurate measurement of the correct proportions of Flexane and the curing agent is a must. Poor results with Flexane are often attributable to carelessness in mixing the correct ratios.

Stir the curing agent before measuring it and adding it to the Flexane. Pour the correct amount of curing agent into the Flexane and stir it thoroughly using a wide blade such as a spatula or putty knife. Do not use a slim rod or screwdriver blade because they will not mix the materials adequately. Mix quickly, but do not whip air bubbles into the mixture. If your finished products have soft spots in them, chances are good that the mixing was incomplete.

Keep moisture out of contact with Flexane. Make sure that all tools, mixing containers, and molds are thoroughly dry before use.

Flexane cures slowly, reaching about 70 percent of its maximum strength in two days at room temperature. It is fully cured in approximately seven days at room temperature. For faster curing, or for curing at low temperatures, accelerators can be used. The manufacturer should be consulted concerning the use of accelerators. Heat will also speed up the curing time. A temperature of not over 150 °F. may be used, in which case curing will essentially be complete in 16 hours.

Chapter 24

Tips on Cutting and Shaping Rubber

Simple rubber parts like gaskets, body cushions, door bumpers, and trunk lid bumpers can be shaped by hand from rubber stock in sheet or slab form. Surplus Flexane may be poured out on a sheet of glass coated with a suitable mold release agent to form sheet stock of about 1/16 inch thickness. Slabs or cylinders of stock may be made by pouring Flexane into suitable molds.

Sheet stock may be cut to form with a sharp knife or a razor blade. Thicker stock can be cut to rough size on a band saw fitted with a coarse tooth blade. If no band saw is available, thick pieces can be cut with a knife if the cut is spread under tension as the cutting proceeds. Wetting the blade with water, soap, or glycerine will help cut thick sections.

DRILLING HOLES IN RUBBER

Holes can be drilled in rubber with ordinary twist drills rotated at high speed. For best results, chill the rubber in a freezer before drilling. Cooling rubber in dry ice is not recommended because the part may shatter when drilled or clamped. Holes made by twist drills will usually come out somewhat smaller than the drill size, so it is good to experiment on a piece of scrap before you drill the final hole.

More precise holes in rubber stock can be made with a steel or brass tube sharpened at one end. It is rotated slowly in a drill press to make the cut. Lubricate the cutter with soap or glycerine.

Holes in rubber stock 1/4 inch or less in thickness should be cut with a leather punch or wad punch.

GRINDING AND POLISHING RUBBER

Rubber parts may be shaped and finished on a disc or belt sander or on a bench grinder. Keep the part in constant motion to avoid burning the surface. For a final, smooth finish, buff the parts on a cloth buffing wheel using rouge as a buffing agent. Wash the parts thoroughly in soap and water to remove the buffing compound.

Appendix

Where to Buy Supplies and Equipment

FOUNDRY SUPPLIES AND MOLDING SAND

Your local foundry supplier will be the best source. Look in the yellow pages of your telephone directory under "Foundry Supplies and Equipment." Petro-Bond products are manufactured by Baroid Chemicals, P. O. Box 1675, Houston, Texas, 77001. You can also contact the following Petro-Bond suppliers:

AMERICAN STEEL & SUPPLY
1011 East 103rd Street
Chicago, Illinois 60628
(312) 928-9893

THE ASBURY GRAPHITE MILLS, INC.
Main Street
Asbury, Warren County
New Jersey 08802
(201) 537-2155

ASHER-MOORE COMPANY
1001 Old Bermuda Hundred Road
P. O. Box 3615
Richmond, Virginia 23234
(804) 748-8123

BALFOUR GUTHRIE (CANADA) Ltd.
740 Nicola Street
Vancouver, British Columbia V6G 2C2
(604) 685-0211

ROBERT A. BARNES, INC.
151 S. Michigan Street
Seattle, Washington 98108
(206) 762-0920

ALEXANDER D. BARCZAK ASSOC., INC.
65 S. W. Tenth Drive
P. O. Box 367
Boca Raton, Florida 33432
(305) 395-3392

BRANDT EQUIPMENT & SUPPLY CO. INC.
2800 N. Nichols
P. O. Box 4384
Fort Worth, Texas 76106
(817) 626-5404

G. W. BRYANT CORE SANDS, INC.
2861 Bryant Road
P. O. Box 133
McConnellsville, New York 13401
(315) 245-1920

THE BUCKEYE PRODUCTS COMPANY
7020 Vine Street
Cincinnati, Ohio 45216
(513) 761-7100

CANFIELD & JOSEPH, INC.
830 Armourdale Parkway
P. O. Box 5035
Kansas City, Kansas 66119
(913) 621-1320

COMBINED SUPPLY & EQUIPMENT CO., INC.
215 Chandler Street
Buffalo, New York 14207
(716) 873-2920

FISCHER SUPPLY COMPANY, INC.
1422 Chestnut Street
Chattanooga, Tennessee 37402
(615) 266-5115

FOUNDRIES MATERIALS COMPANY
5 Preston Street
P. O. Box 336
Coldwater, Michigan 49036
(517) 278-5668

FOUNDRIES MATERIALS COMPANY
8951 Schaefer
Detroit, Michigan 48228
(313) 933-6444

JOHN M. GLASS COMPANY INC.
648 S. East Street
Indianapolis, Indiana 46225
(317) 635-5131

HAMILTON-PARKER COMPANY
491 Kilbourne Street
P. O. Box 4086, Station H
Columbus, Ohio 43215
(614) 221-6593

HOFFMAN FOUNDRY SUPPLY COMPANY
1193 Main Avenue
Cleveland, Ohio 44113
(216) 241-4350

INDEPENDENT FOUNDRY SUPPLY COMPANY
6463 E. Canning Street
Los Angeles, California 90040
(213) 723-3266

INDUSTRIAL & FOUNDRY SUPPLY COMPANY
2401 Poplar Street
Oakland, California 94607
(415) 452-1226

H. R. JONES TOOL & SUPPLY COMPANY, INC.
1710 Spring Garden Street
P. O. Box 20023
Greensboro, North Carolina 27420
(919) 275-0491

KIMBALL CHEMICAL COMPANY, INC.
Old Sinclair Yard
P. O. Box 880
Sand Springs, Oklahoma 74063
(918) 245-6666

KLEIN-FARRIS COMPANY INC.
531 Statler Office Bldg.
22 Providence Street
Boston, Massachusetts 02116
(617) 426-1210

LA GRAND INDUSTRIAL SUPPLY COMPANY
2620 S. W. First Avenue
P. O. Box 8053
Portland, Oregon 97207
(503) 224-5800

LARPEN SUPPLY COMPANY, INC.
5501 West State Street
Milwaukee, Wisconsin 53208
(414) 771-4760

THE MARTHENS COMPANY
204 38th Street
Moline, Illinois 61265
(309) 762-5586

MEDINA ART CASTINGS LTD.
The Marine Villa
Undercliff Drive
Ventnor
Isle of Wight
PO38 1UW
England
Tel.: 0983-855822
Fax.: 0983-852146

MIDVALE MINING & MFG. COMPANY
6310 Knox Industrial Drive
St. Louis, Missouri 63139
(314) 647-5604

MONINGER FOUNDRY SUPPLIES, INC.
7330 W. Cortland Street
Elmwood Park, Illinois 60635
(312) 453-4240

NORTH AMERICAN MINERALS COMPANY
Division of J. C. Keaney & Sons, Inc.
101 Pennsylvania Boulevard
Pittsburgh, Pennsylvania 15228
(412) 563-1234

PENNSYLVANIA FOUNDRY SUPPLY & SAND CO.
6801 State Rd.
Building "B"
Philadelphia, Pennsylvania 19135
(215) 333-1155

THE PERRY SUPPLY COMPANY, INC.
831 First Avenue North
P. O. Box 1237
Birmingham, Alabama 35201
(205) 252-3107

SMITH-SHARPE COMPANY
117 27th Avenue S.E.
Minneapolis, Minnesota 55414
(612) 331-1345

STOLLER CHEMICAL & FOUNDRY SUPPLY, INC.
352 Mill Road
P. O. Box 418
Wadsworth, Ohio 44281
(216) 336-6628

VAL ROYAL LaSALLE LTD
12200 Blvd. Laurentien
Montreal, Quebec H 4 K 1 N 2
Canada
(514) 227-2101

THERMOCOUPLES

You may be able to obtain thermocouples from firms that sell or repair industrial furnaces. If you cannot find a local source, you can purchase them from the address below. Specify chromel-alumel, 8-gauge, 18 inches or more, with ceramic insulators. The cost is about $9-10 each, including shipping charges. Minimum order is $50.

California Alloy Company
1475 Potrero
So. El Monte, California 91733
(213) 579-3230

About the Author

W.A. "Bill" Cannon is the technical editor of *Skinned Knuckles*, a monthly magazine devoted to the restoration, operation, and maintenance of all collector vehicles. He was an engineer and scientist in the chemical, automobile, and aerospace industries for thirty years before he retired.

He claims a life-long interest in all things mechanical, but his work as an engineer for the Ford Motor Company in the 1950s sparked his interest in old cars. He spent many a lunch hour in Dearborn, Michigan, surveying the Ford Museum's car collection housed in a building adjoining the company's engineering offices.

He is a native of the State of Washington, and served in World War II.

Index

Index